The Lambretta people

By Stuart Owen

SNOOP MEDIA

Dedicated to the memory of Pete Meads

Acknowledgements

Thanks to the following people who have contributed to the making and
production of this book
Anthony Tessier, Joan Todd, Maurice Knight, Pete Meads, Peter Nottingham,
and Sheila Meads
Special thanks go to Bob Wilkinson for his time, help and dedication

Contents

Introduction

Forward by Bob Wilkinson

Introduction

There are moments in history when something so creative and ingenious comes along that it leaves a lasting effect. The Lambretta motor scooter was one of those creations not only serving millions of people during its time of production but leaving a legacy that continues to this day. While Italy may have been the nation that gave us the Lambretta, Great Britain was most definitely its second home. A classic design it may have been, but without certain events and people, it may never have happened that way.

This book is dedicated to them telling the story of the struggles, the success, and the heartache as one of Italy's finest marks took hold of the British public. Key figures who were responsible for creating a legacy like no other, they are the Lambretta people.

Forward

*T*o the man who incredibly drove a Lambretta at over 100mph and once called me Mr. Lambretta on a very personal occasion. However, in all honesty, I can think of dozens of other very special people who deserve this accolade far more than I do.

I remember to this day many well-known and internationally famous members of the Lambretta Club of Great Britain. Members who in the Swinging Sixties spent most weekends enthusiastically and passionately demonstrating their driving skills and professionalism during rallies, endurance trials, hill climbs and many other sporting and competitive activities. All are responsible for building the camaraderie into the nationwide Lambretta family we are so proud to be part of today.

What an amazingly remarkable life we have enjoyed and continue to savour simply due to these incredible Italian treasures initially created from misshaped scaffold poles in a Milan village.

The Lambretta legend will last forever.

Bob Wilkinson

Chapter One

The speedway riders

*R*ecovery wouldn't happen overnight; at best, it would be at least a decade before life would return to some sort of normality. That was the view of all European countries, counting the cost of the damage WW2 had bestowed upon them. Britain had suffered enough damage, but the country's people were prepared to rebuild it and manufacturing would be at the heart of the recovery. Many businesses had concentrated on the war effort but were ready to thrive once again, and with so many resources needed, those with an entrepreneurial spirit had the potential to benefit the most.

Les Ashton had left the RAF at the end of 1945 with nothing to his name, just a few meagre belongings, but he was determined to succeed. Now a veteran dressed in his demob suit, he returned to the capital to start rebuilding his life. Living in Wimbledon, he was known and regarded as a mischievous character, a wheeler-dealer who would turn his hands to almost anything to make a fast buck. It wasn't anything dodgy or illegal, perhaps involving a cash deal at best; it was just his way of dealing with day-to-life.

So, it seemed quite a shock that this man soon had a thriving business not only with a shop front but also several lockups. No one knows how it happened, but by 1948 Les Ashton was the proprietor of Rapid Motors, a substantial double-fronted motorcycle dealership based in Haydons Road Wimbledon. At the time, getting a dealership was no mean feat and becoming an agent was even more difficult. The shop happily traded with B.S.A, Velocette, Vincent, and many other well-known motorcycle manufacturers of the day. The impressive list of names made it appealing to the general public, desperate for any form of transport and with motor cars out of most people's reach and petrol still being rationed, these were good times for any motorcycle dealer.

Les pictured in his demob suit after leaving the RAF circa 1946, standing next to his mother

Being associated with motorcycles opened up a whole new world, and for Les, one of those was speedway. Post-war, it became a vast sport and pastime for those competing and the crowds who came to watch in their thousands. Bar football it was the biggest spectator sport out there; in each town and city, it seemed there was a team, and in London, more so. Les was actively involved with the Wimbledon speedway club known as "the Dons" and was responsible for tuning the engines and setting up the bikes for the riders. He soon became involved with some of those riders and housed Ronnie Moore for a while, who would later become world champion, so much was his passion for it all.

The club boasted other top riders competing internationally, representing the England team across Europe and Les, alongside fellow motorcycle dealer Donato Royale became the mechanics on many of their trips abroad. Donato (Don) had been a competitive rider before the war and though now retired, still played a significant role in the team, having gained so much experience. International duty saw them travel far and wide, which on several occasions would take them to Italy. It was here that they would make a life-changing discovery that would forever alter the course of two-wheeled travel in Britain.

While taking time out before the race meetings, they couldn't help but notice these small two-wheeled, two-stroke machines buzzing around the streets of Italy's capital Milan. A far cry from the much more prominent agricultural motorcycles they were used to dealing with. While there was no direct comparison, an air of curiosity made them investigate this new phenomenon further. There were two main choices: the Vespa made by Piaggio or the Lambretta built by Innocenti, which was based on the capital's outskirts. The latter began to attract their attention, and upon returning home from one of their excursions, they made further enquiries.

It soon became apparent that the Lambretta had already arrived in Britain by way of a company called the Anglo Italian Trading Company which had imported a couple of dozen model B examples in the summer of 1948. Evidence of this came by way of a review of it in "Motor Cycling" magazine, but it had done very little if anything to help sales which were virtually nil. Les and Don saw potential, from its low price and economical engine to its step-through design which could appeal to anyone, male or female. It was at least worth a punt, and Les, with his gung-ho attitude, decided they should try and see if it were possible to stock it.

There were rumours of others trying out the same idea, with another speedway rider Wilf Plant bringing a couple of examples back from Italy while on international duty. He, too, had a motorcycle dealership and, having sold them both, made enquiries to Innocenti regarding the concession. Ultimately, he shied away as he didn't have the funds to take the quantity the company asked him to commit to if he wanted the sole import rights. Whether or not Les knew about this is unknown, but he soon acted upon the chance realising others were sniffing around the idea. Along with Don and a man named Egidio Rosso, the three of them approached the Anglo-Italian Trading Company about taking over the import of the Lambretta. The company was going nowhere with it and didn't have the trade connections they did, so they soon struck a deal to take it off their hands.

DUPLICATE FOR THE FILE.

No. 489574

Certificate of Incorporation

I Hereby Certify, That

CONTINENTAL MOTOR CYCLES LIMITED

is this day Incorporated under the Companies Act, 1948, and that the Company is Limited.

Given under my hand at London this Nineteenth day of December One Thousand Nine Hundred and Fifty,

Registrar of Companies.

Certificate received by

Date 19/12/50

The company was initially formed in the last month of 1950 to import the Lambretta and sell it in Britain, having agreed on a deal with the existing importer who found it hard going trying to sell it

The three men set up a new company, Continental Motorcycles Limited, which would be the sole concession for the Lambretta in Britain. 4th December 1950 would be the official company registration date with a total capital of £1000 introduced into the company. It was divided into shares worth £1.00 each, with all three men having an equal amount to allow it to be fair. The trading address would be at Rapid Motors, which belonged to Les but remained a separate company, allowing him to import and sell them at the same premises. Don, whose shop was based in Aldershot, would naturally be the other location that stocked the Lambretta, meaning both would benefit from this agreement. Egidio seemed happy with this deal as he would make a profit from Continental Motorcycles limited the more the other two could sell them.

It was a daunting challenge with the Lambretta virtually unknown in Britain and very little by way of exposure other than the motorcycle press, which so far had been vague with its coverage. The main problem was the motorcycle manufacturers who dominated the industry and, to a certain extent, governed the input into any magazine as they advertised heavily with them. Any outside threat was heavily frowned upon even though the motor scooter concept was regarded as a joke or a quick fad at best, something they need not worry or bother about, even if it got sporadic coverage. When it did, it was usually only a few lines and a blurry picture that seemed hardly worth mentioning.

The only surviving company registration certificate logged the three men down as directors; however, the document has the names of Don and Egidio crossed out, presumably after they had transferred their shares to the Agg's

There was help on hand for the fledgling company by way of Innocenti, who had grand plans for the Lambretta and carried out many publicity stunts to attract attention. Part of that was speed record attempts, which were a significant way of gaining media exposure if setting records. While winning the Isle of Man TT and Grand Prix meetings was respected, claiming to be the fastest motorcycle carried a considerable amount of kudos and made the public take notice. With their tiny 125cc engines, Innocenti joined in even though their torpedo-shaped machines designed to cut through the air looked like anything but a Lambretta. The British magazines of the time widely reported on the speed record meetings and were soon forced into announcing the Lambretta on more than one occasion. The 125cc class became heavily contested as many manufacturers based this size engine on their mainstream machines, and no one thought the Lambretta would stand a chance. However, Innocenti repeatedly proved that they were wrong in their thinking by breaking records almost monthly.

Again that Lambretta !

NOT content with their astonishing series of successes obtained recently at Montlhéry, the Lambretta team, Macerini, Bruori, Masetti and Rizi, got busy again, last week, on the French speedway. Riding the same 125 c.c. Lambretta, they broke their own 24-hour record, set up on March 23-24 last, by a handsome margin: 2,449.631 kiloms. were covered in the 24 hours, representing an average of 102.067 k.p.h. (63.75 m.p.h.). as against their previous 94.517 (59.05 m.p.h.).

As it is less than one month since the previous figures were established, this latest 24-hour achievement does not constitute a world's record.

Continuing after this for more than another 24 hours, the team made the following records, which are, of course, subject to official confirmation: 3,000 km. in 30 hrs., 32 mins. 15.25 secs. (98.239 k.p.h.) (61.006 m.p.h.); 2,000 miles in 33 hrs. 1 min. 49.75 secs. (97.445 k.p.h.) (60.513 m.p.h.); 4,000 km. in 41 hrs. 7 mins. 14.3 secs. (97.274 k.p.h.) (60.407 m.p.h.). In 48 hours, 4,686.417 kiloms. were covered (97.639 k.p.h.) (60.634 m.p.h.); 3,000 miles in 49 hrs. 23 mins. .12 sec. (97.759 k.p.h.) (60.708 m.p.h.); 5,000 kiloms. in 51 hrs. 8 mins. 3.95 secs. (97.781 k.p.h.) (60.722 m.p.h.)

MORE LAMBRETTA RECORDS

WITH racers Ambrosini, Ferri, Masetti and Rizzi in turn in the saddle, a fully streamlined 123 c.c. two-stroke Lambretta scooter was ridden round Montlhéry at intervals between September 27 and October 5, to add several more world's records to those already held by this remarkable Italian machine. When the latest figures are officially confirmed by the F.I.M. it will mean that all but six of the 36 recognized 125 c.c. class records are in the name of Lambretta. The following speeds were achieved:—

10 km., 83 m.p.h.; 10 mile, 85 m.p.h.; 50 km., 87.5 m.p.h.; 50 mile, 88 m.p.h.; 100 km., 88 m.p.h.; 100 mile. 88.5 m.p.h.; 500 km., 85.3 m.p.h.; 500 mile, 82.8 m.p.h.; 1,000 km., 82.4 m.p.h.; 1,000 mile, 82.4 m.p.h.; 1 hr., 88.5 m.p.h.; 2 hr., 88 m.p.h.; 3 hr., 86.5 m.p.h.; 4 hr., 85.5 m.p.h.; 5 hr., 83.1 m.p.h.; 6 hr., 82.7 m.p.h.; 7 hr., 82.3 m.p.h.; 8 hr., 82.5 m.p.h.; 9 hr., 82.3 m.p.h.; 10 hr., 82.4 m.p.h.; 11 hr., 82.2 m.p.h.; 12 hr., 82.4 m.p.h.

Innocenti was constantly establishing speed and endurance records with the Lambretta during 1950-51, which were widely reported in the British press and gave vital coverage.

The name Lambretta beginning to crop up and associated with success must have affected the public, leading to enquiries for road-going machines. Slowly the Lambretta started to sell, and not just through the two shops belonging to Les and Don, but other dealers prepared to become an agent. Les was excited about the whole thing, and his daughter Joan remembers when the first shipment came in. "Though the shop was in Wimbledon, for some reason, they landed in Twickenham, so my mother, father, and I went over in the motorcycle sidecar combination to have a look" "There was a couple of dozen or so, and dad couldn't wait to unpack them he was so happy they had arrived". It was only the tiny shoots of what was to become, but as sales grew, it created a knock-on effect that would cause a big problem for the three men.

While there was profit to be made from selling the Lambretta, the costs of running the business and wages meant much of it was eaten up, meaning there wasn't much left over to help grow the company. There is no record of how many machines had been sold in the early days, but by the middle of 1951, just six months after starting the company, they needed to order more from Innocenti. That's fine; repeat business getting them shipped over, stocking the showrooms, and selling them was the way to success. The problem was cash flow, as Innocenti wanted payment upfront, which was standard practice in the industry, but without sufficient funds to pay for a big order, it would be a struggle. Les realised they were on to something good, and given the correct exposure over time, the Lambretta could be a big seller, and there was no way he would let this opportunity slip through his hands. They needed a financial backer, someone who could give them the cash injection they needed to continue.

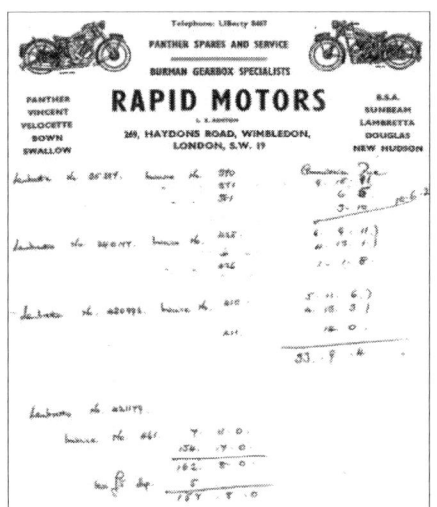

The trading address for Continental Motorcycles was at 269 Haydons Road, Wimbledon, where Rapid Motors was based and owned by Les. Continental Motorcycles would then sell the imported machines at trade price to Les and Don, who then sold them to the public through their own companies. Therefore, they would benefit twice, whereas, for Egidio, only once

The problem with that idea is that the backer would want to see a return and sometimes very quickly, depending on the conditions. If this were the case, it would put them under a lot of pressure to get sales moving even faster, which wasn't guaranteed as the Lambretta idea was still in its infancy. Another thing that stood out was if the new investor wanted to start making decisions on how they ran the business. Les and Don knew the trade and how to work it; interference from anyone else was something they could do without. Regardless of the potential problems, this was the route they chose to go down and little did they know how it would change things in the future. The question that needed answering was how do you find someone with the funds and prove to them it would be a worthwhile investment?

The answer came through Don, who knew someone that could potentially fit the bill. However, it wasn't as clear-cut as it first seemed, as Don was in financial trouble with his own business. He owed money and needed to pay it off before asking for more funding. It came to light that it may have come from the same person they were about to ask now and that it wasn't, in fact, his own money he put into Continental Motorcycles Limited when they set it up in the first place. No factual evidence was written down, but by June 1951, significant changes were made to the company. Egidio Rosso was removed as a shareholder, as was Don Reale and in their place were two new names, a father and son, James Agg senior and his son Peter Agg. James Agg had been involved in banking and, for a while, wine importation and was well addressed when it came to acquiring goods from overseas. Peter Agg had been dabbling in the car trade but showed little interest in scooters; he was more akin to sports cars and racing. Mr Agg senior was retired but agreed to come into the new venture while his son, a keen up-and-coming businessman, saw potential in the Lambretta. It was strange that Egidio Rosso was replaced without any explanation, only for him to turn up in later years when the Lambretta empire was at its strongest as controller of the company's technical catalogues and pricing guides.

As for Don, his financial troubles were too big, and he could not pay the Agg's the money he owed them; they took his debt on by taking his shares over in the company and, judging by the amount they acquired, bought out Egidio at the same time. Joan never saw Peter Agg but remembers James Agg coming to their house several times. She was too young to understand what they were discussing but realised it was something to do with the business. This was confirmed when he visited one last time with a bottle of wine to celebrate them doing a deal, vividly etched in her mind because Les was tea total and so didn't drink a drop.

All this was confirmed in June 1951 when a share certificate was issued, but it showed a considerable change and one that would have a significant impact. With 99 shares to his name, Les Ashton was now only a minority holder in the company compared to the 615 the Agg's now held. It was a remarkable turn of events as the men who had brought the Lambretta to Britain were no longer in charge of the company they had started, all in less than six months. Those now in charge had no connection with it whatsoever, but perhaps that was a good thing. They soon realised the Lambretta was virtually unknown and, as they were not from a motorcycle background which British manufacturers dominated, wouldn't be scared by them. As far as they were concerned, the Lambretta was a commodity, and with hard work and dedication, it may be possible to make a success of selling it.

The new share certificate shows that the Agg's now had overall control of the company, with only Les remaining as one of the original three. It was a big decision for him to make, not only letting two people he didn't know to buy into the company but having a bigger say than him. Only time would tell if this would be a move he would come to regret

THE COMPANIES ACT, 1948

COMPANY LIMITED BY SHARES

Special Resolution

– of –

CONTINENTAL MOTOR CYCLES LIMITED.

Passed the 17th day of August 1951.

At a GENERAL MEETING of the Members of the said Company duly convened and held at the Registered Offices of the said Company, 64, High Street, Epsom, Surrey, on the 17th day of August 1951 the following SPECIAL RESOLUTION was duly passed :–

"THAT the name of the Company be changed to LAMBRETTA CONCESSIONAIRES LIMITED." ——

Chairman.

WE CERTIFY that this Resolution has been printed by the Lithographic process.

For H. A. JUST & Co;

A special resolution was granted to change the name to Lambretta Concessionaires limited, which was far more appealing. It was undoubtedly the most significant decision made in the company's history. It is also noted that the registered office was now 64 High Street, Epsom

To do so would mean full cooperation between Les Ashton and the Agg's because any disagreements would give them less chance of succeeding. At the time, Innocenti was producing the open-framed model C and the enclosed LC, the forerunner to the LD. These would be the models they would concentrate on, and with proven reliability and economical to run, that would be the focus of attention when advertising them. The Vespa, which was already proving to be the main rival, was made under licence by Douglas of Bristol, and they were having some reliability issues, mainly with electrics. Focusing on the reliability of the Lambretta was vital, and more importantly, any issues with breakdowns should be attended to immediately, putting customer care at the top of the list.

For now, just selling them was the priority, and to do so would mean advertising even though only limited funds were available. Peter Agg had seen how well the Vespa was doing, and this was down to the near blanket coverage in the automotive press. Douglas was a big company that had been producing motorcycles for years and had the budget to do this, not forgetting Piaggio backed them in Italy. The three-person band of Continental Motorcycles Limited had far less, and even though the Agg's had money behind them, how much did they dare throw at it?

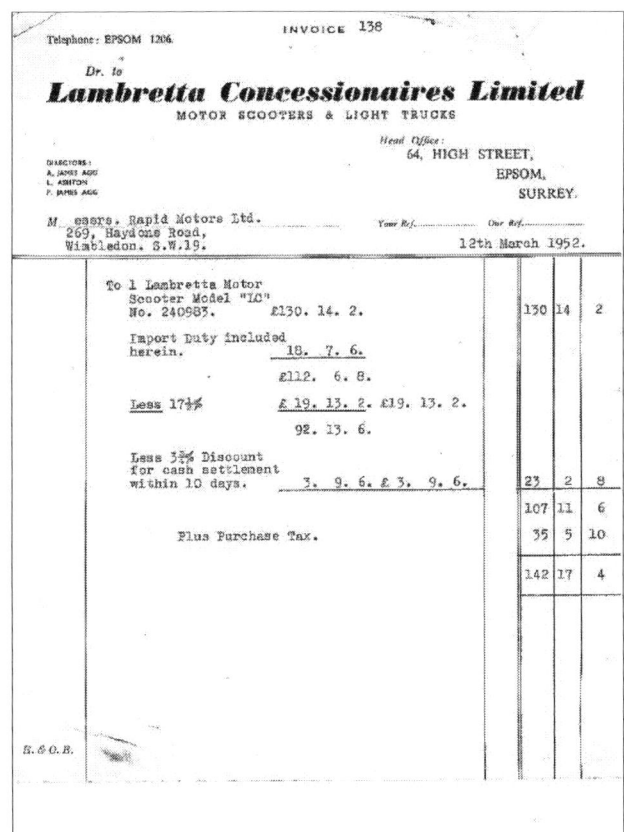

The move to Epsom is shown on this early invoice dated March 1952. It clearly states that both the Agg's and Les Ashton are directors. There were many invoices to Rapid motors, so like when the company first started, Les was benefiting twice from the deal as an importer and retailer

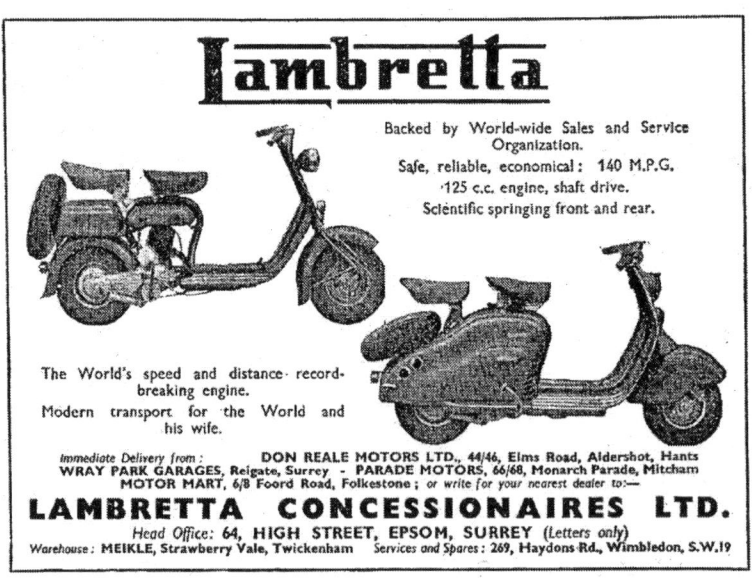

One of the earliest advertisements lists the same address but mentions that servicing was still carried out at Haydons road. More important is the fact that there are now four shops listed as agents as the dealer network began to grow

One thing that stood out was the company name, which didn't mention the Lambretta but motorcycles, a different concept from the scooter. As far as Peter was concerned, the name needed changing, so in September 1951, the company was renamed Lambretta Concessionaires Limited. For a while, they still traded in Haydons road again; this could be off-putting because Rapid Motors was a motorcycle shop and having the Lambretta sandwiched between bulky motorcycles didn't portray the right image.

That would mean a move to new premises at 64 high street, Epsom, with the shop front and sign dedicated purely to the Lambretta. Servicing would remain in Haydons road, but that would now be its only capacity within the business. There were disagreements between Les and Peter, and they could sometimes be fiery, but it was to be expected. It was hard for Les as he was the only one of the original three who remained, and the Agg's had, in effect, taken over all decision-making and were in complete control. It couldn't continue like this much longer; by October 1953, Les had been pushed out. There were all sorts of rumours about bust-ups and arguments, but the outcome was that the Agg's now owned all the shares between them. The fallout between Les and Peter was very acrimonious, so much that years later, the only reference Peter Agg made to the time before he became involved was that the business had been based in Haydons road for a short while. Les felt aggrieved as it was him that had made it possible for this to happen and then been written out of the story.

Les sitting on a Panther, one of the brands he continued with at Rapid Motors after leaving Lambretta Concessionaires

He continued with Rapid Motors for years to come and took on other scooter concessions, including the French-made Terrot. In adverts for it, he would boldly state that he introduced the Lambretta to Britain, but it was all in vain. The Agg's had wrestled control of the Lambretta from him, and for the speedway riders, this is where their part in the story ceased. Regardless Les Ashton must always be remembered as the person who began the British Lambretta legacy and must never be forgotten for what he did.

Chapter Two

The smart gent and the AA man

*B*y the time Peter Agg was in control of Lambretta Concessionaires, he was just into his 20,s and armed with a potentially colossal selling product, keen to make sure it succeeded. Always smart in appearance, dressed in a suit and sporting a handlebar moustache, he acquired the city gent look that oozed class. His real passion was sports cars, and he was keen to enter races where possible, but to do so required significant funding. He believed the Lambretta could be the key to his dream coming true but realised it wouldn't happen overnight. Quite the opposite: building up a profitable company takes years with a lot of hard work and effort. He was determined to succeed and prepared to do whatever was required to make his dream a reality, so to him, the small Italian motor scooter was just a commodity. Still, by 1953 it hadn't generated the revenue to allow him to pursue the lifestyle he craved, and something needed to happen to make the breakthrough.

His vision for the Lambretta was bold and would mean taking on the might of the two-wheeled industry that big British manufacturers dominated. Perhaps that's why the Lambretta hadn't been taken up earlier, as no one gave it a chance against the might of BSA, Triumph, Norton and a whole host of others. While they were strong opposition, that wouldn't be the most brutal battle; that accolade went to the public as virtually no one had seen or heard of a scooter before 1950 in Britain, certainly not mainstream.

The Vespa may have had some slight success, but in the grand scheme, its sales paled into significance compared to motorcycles. These would be the glory days of two-wheeled travel as the car was only affordable by the more affluent in society and was beyond the reach of most households.

Peter Agg always smart in appearance and sporting his trademark handlebar moustache

The motorcycle didn't offer the same weather protection and less carrying capacity, but those surviving on low post-war income saw it as a necessary form of transport. If Peter Agg was going to stand any chance of making the Lambretta a success, it had to start right now while two-wheeled vehicles were in their prime. Undoubtedly, he had a vision of where he wanted the company to be and knew it would be possible if he implemented the correct plan. In the beginning, there was only six staff, and one of them, James Agg senior, only played a financial role, leaving the business's day-to-day running to his son and the employees.

Plans were drawn up to create a dealer network and, over time, make it national. Once up and running, it would be backed up by service centres with the idea that the customer could buy in confidence, knowing they were being looked after once they had made a purchase. Nothing like this had been attempted before in the automotive industry because it would be almost impossible to implement. Peter was young and full of drive and enthusiasm; there was no stopping him, so without hesitation, he began to put his idea into motion and started recruiting. Finding labourers to uncreate and dispatch machines was easy; getting those with the necessary business skills to sell the product was not so.

Maurice Knight started working for the AA in 1952 in their information department on general motoring knowledge. It was a rather mundane job, but he persevered even though he didn't enjoy it. The pay was fair, and with it giving him financial security, he decided to stick with it until a better offer came along. By now, the Lambretta had come to the attention of both the RAC and AA; seeing its popularity increase, they decided they must cater for its needs. Shortly afterwards, an AA customer contacted the company directly regarding the hire of one and how to go about it. It was a strange request as their job was to help people with breakdowns but to put the customer's needs first, they would make enquiries. Maurice was tasked with sorting the request and

being one of those who got on with the job; his solution was to ring Lambretta Concessionaires directly to find out the answer.

Maurice pictured with Edna Calder, the company secretary at Lambretta Concessionaires

It was a quick telephone conversation and straight to the point, but one that would change his life forever. By chance, it was Peter Agg who answered the phone, and upon hearing the request, they began a lengthy conversation, talking about the two-wheeled industry. Neither man had spoken to the other before, but Peter was impressed by what Maurice had to say and invited him to the factory at Wimbledon. Surprised he might have been, but perhaps there might be a job opportunity and one that could get him out of the mundane and rather dull routine at the AA, so he took up the invitation. Peter seeing potential, was, going to offer Maurice a job as a technical representative; however, by the time he got around to visiting, it had already gone to someone else, a coach driver who thought he knew his way around the mechanics of any vehicle. Peter told Maurice not to worry as he felt the demand was getting bigger and explained, "we can have two technical representatives instead of one". Shortly after, the coach driver, who wasn't as gifted as he made out, left, meaning Maurice, who had openly accepted the job, was the sole person in charge of that department.

It was far better than being at the AA, where you were one of many employees compared to Lambretta Concessionaires. There was only a handful, and Peter gave him the freedom to learn and grow into the role. Even so, he instructed Maurice to take a model C and D, which had recently been introduced and completely strip them down to understand all their workings. Three months were spent locked away in a small workshop, and by the time he had finished, Maurice knew where every single nut and bold went on each machine. The idea was he could now go out to the growing list of dealers and teach them the mechanics of the Lambretta's they were selling. Even though he had changed his job, it still meant being out on the road travelling across the country, not that it mattered as he thoroughly enjoyed what he was doing by this time.

Peter Agg had become annoyed that the approved dealers weren't buying the factory tools required to do repairs, certainly regarding the engine. Maurice explained that if they don't know how to work on them, what's the point of them buying equipment, they won't use? Telling him not to worry, Maurice went on a purge and started training mechanics at each shop he visited, and in doing so, orders came in for tools. Peter soon realised that left to his own devices, Maurice would not only do his job but also improvise and improve the company.

Peter outside the shop in Epsom

He recalls going to Horner's in Manchester using an old Morris van with a model D thrown in the back. "It was a long way from London, and the clutch was tough work" "when I got there, they had no workspace, so I ended up stripping it down on the pavement in front of the shop". Some days he thought was it worth all the effort but could feel they were on to something big and wanted to be part of it. Peter Agg began to notice another talent in Maurice, which came about from the factory tool saga, that of a salesman, and it didn't take long for him to take advantage of it.

Earning £7.10s a week wasn't a bad wage for Maurice to take home, but there was a chance of a promotion to a sales rep. The wages were lower, £5.00 a week, but another £5.00 for each dealer he could sign with a shilling for every Lambretta sold. Though that would mean a drop in wages, the commission gave him the incentive to help build up the network of dealers. The more he did, the more machines they would sell, so the commission would far outweigh his previous salary. Peter Agg was clever because he knew Maurice would have to get results to earn a good wage, but he had been given plenty of financial encouragement with the package on offer. He could tell early on that Maurice would be good at his job and that it would help build up the Lambretta empire much quicker.

As the company embarked on an extensive advertising campaign, the Lambretta became more noticed, and sales picked up tremendously. Combined with the ever-growing service agent scheme, they offered a package to the customer no other manufacturer could compete with

It would still mean being out on the road, but this time, not getting his hands dirty, instead wearing a smart suit giving it the big sell. Unfortunately, driving the Morris van was still the method used in getting around to each dealer, but if things went right in time, he was assured it could be replaced with something much better, again an incentive to do well. Life on the road could be challenging at the best of times; Maurice remembers trying to sell a bubble car when they ventured into that fad for a short while. It caught fire during the demonstration with him inside, luckily managing to escape and still making a sale. He would often pull up at a shop hoping to get them to sign up to the Lambretta, only to find the Vespa rep already there. "I would bump into them every week, and while some were okay, others would simply call me a bastard, making me even more determined to get the signature".

There were many towns and cities across the country, and Peter Agg had made the ambitious target of getting a Lambretta dealer in most of them. With the commission now in place, each time Maurice got a dealer signed, he was hungrier than ever to succeed. As the 1950s rolled on and the Lambretta name became more widely known and respected, it was easy pickings. It wasn't like someone had to set up a business to start trading them as they were explicitly targeting existing motorcycle dealerships, in a way, a preprepared network. So, what if rows of big, black-painted British motorcycles were lined up in a showroom? A brightly coloured row of Lambretta's next to them would brighten the place up. It must be remembered that many of these dealers had huge showrooms with extensive window displays, so the idea was to convince them that they could cater for both types of vehicles by displaying them side by side. The significant advantage a Lambretta, or any other scooter for that fact, had over a motorcycle was that anyone could ride one. There was no petrol tank to straddle over or a big oily chain to get in the way, and combined with better weather protection; they appealed to women even more so. Maurice would reiterate that this was one of the top sales points as the dealer realised that by stocking the Lambretta, there was a substantial new trading potential that had previously been unobtainable. That's why it was easy to sign them up, but there was a strict criteria of rules they must adhere to.

What made it difficult was that many of those he visited had already tried the Vespa and its electrical problems made them very wary about the motor scooter. At the very beginning, there were just six dealers on board, and Maurice went out every day hoping to sign a new one. Try though he might, there were weeks where not a single signature was obtained, and it was starting to worry him. Peter became slightly concerned that maybe Maurice was struggling, so he set up a test where he and Maurice took a van each with twelve Lambretta's loaded up inside. It was a sportsman's bet, and the first one to return with an empty van was declared the winner. Maurice targeted Comerford's, a big dealership and had returned with an empty van by midday. For Peter, it took much longer, another 24 hours, and that's when he was sure he had the right man for the job. There were bound to be weeks where sales were poor, but this confidence-boosting exercise had worked perfectly.

The mobile units were another idea that allowed dealers and owners to experience first-hand the backup on offer

To ensure confidence in a dealer, Peter and Maurice devised an idea to protect them and give them more security. If the dealer invested heavily in machines, spares, and workshop tools, it would be no good if they had other competition in the same town. In places like London and the big cities, there would be several dealers as it was a more extensive marketplace and catchment area, but that principle wouldn't work in smaller towns. Maurice had approached Harrisons, the leading motorcycle dealer in High Wycombe; they weren't interested, mocking the Lambretta and almost pushing him out of the door. Undeterred, he visited a small family-run business in the same town, Desborough Cycle Works, which was more than happy to take on a dealership and began selling large quantities of Lambretta's. Harrisons were soon back on the phone pleading with Maurice for the right to sell them, but true to his word, he declined them to ensure the existing dealer had protection in their investment. It was a policy that any dealer who showed loyalty to them had it reciprocated and made it much easier to sign them up in the first place.

Dealers had to meet specific targets to ensure they were pulling their weight regarding sales, as their contract was renewed every twelve months. With plenty of others waiting in the wings, if they didn't, there was a certain degree of pressure to make sure they delivered. However, as the Lambretta began to take off and dealers sold more significant quantities, Peter Agg wanted even more improvement. A lot depended on where they were situated, either in a big city or a rural town, but whatever the sales figure for that year had been, he would double the target for the following one. Maurice said that was an impossible figure to achieve no matter how popular the Lambretta was becoming, and for the first time, they had quite a disagreement on how they ran the business.

Maurice was the one out on the road listening to what dealers said, with most airing their concern that it would put them under too much pressure to reach such high targets. Peter began to think the dealers were controlling Maurice, so he went to see for himself, got the same response, and had to agree and back down on his demands. From now on, as long as there was improvement year on year, that was enough to be content with. It also showed that while Maurice wanted them to sell more Lambretta's, he was realistic about what could be achieved. He became the negotiator between the two sides of the business, trying to see reason and find a workable solution.

The disagreement hadn't harmed and, if anything, made a stronger bond between the two men. Maurice had a better idea of how to go about what Peter was trying to achieve and used the same scheme he was offered when first aggreging to be a salesman. Instead of demanding sales figures must improve, why not offer the dealers a better earnings potential the more they sell? In other words, give them the incentive to make more money if they can reach specific figures. This could be with extra discount on bulk orders of machines and bonus schemes. The dealer would make more money but so would Lambretta Concessionaires, even if the margins were slightly finer. The shop salesmen would, in return, most likely get a bonus from the owner, and as they made the sales, this was a vital aspect of making the scheme work.

As more and more Lambretta's were sold, ownership started to become a way of life and not just a form of transport. This opened up the potential for another form of revenue by having an owner's club

Also, those shops that did perform better would get priority when a new model was introduced. Peter Agg was spending a significant budget on national advertising, and knowing in advance when Innocenti would be bringing out a new model, he would promote it heavily. Dealers would be inundated with enquiries, and those that did get priority could sell it upfront, knowing they would get them first. What began to happen was a more precise understanding between all sides and everyone being rewarded for working as a team. No one worked it all out by themselves, but Peter began to realise more and more just how good Maurice was and what an asset he was becoming to the company.

The LD model had been a great servant and was responsible for making the Lambretta the world leader in scooter production. However, it needed to be updated, and when the Li series was launched in 1958, it was a huge step up in performance and comfort. For Peter and Maurice, it was music to their ears because they knew the public would love it. They were soon proved right as Innocenti struggled to keep up with demand because so many were being sold in Britain

There was an instance when they were trying to sell the three-wheeler van, which had been an uphill struggle. It was the FD 150 type before the Lambro van, basically a shaft-driven model D with a broader chassis at the back housing and open storage compartment. The small power unit meant it wasn't up to long journeys and was only suited to large factories and industrial estates. Peter Agg loved it, whereas Maurice didn't; he hated it and thought it had limited potential. He explained that dealers weren't keen on it either taking up too much space in the showroom, but Peter wouldn't listen. Angry by the comments, he decided to prove them wrong and began touting the idea around big companies that he thought would benefit from such a machine.

Then a breakthrough as the directors of British Overseas Airways Corporation, or BOAC for short, expressed an interest. They thought it would be ideal for ferrying the luggage and freight to and from the planes. Peter was confident he had the deal in the bag, but there was a problem. The baggage handlers didn't like the machine and thought it was dangerous. He was incensed and immediately summoned Maurice to go and do a demonstration of how good it was. Maurice had that sort of why me look about him and, if he was honest, agreed with what the baggage handlers said. "Don't come back without a sale, Maurice", were Peter's last words as he worriedly left the office.

The scene was set with the plane on the runway and all staff and directors present. Maurice drove the FD towards the plane, oblivious that the engines were still running. As he did, the force from the propellers blew the tiny Lambretta van over, throwing Maurice out in the process. If there was a deal beforehand, there certainly wasn't one afterwards, but Maurice felt vindicated that he was right even though he was a little bruised. It was another instance where Peter Agg should have listened in the first place. While it wouldn't be the last time they would disagree, it proved that when Maurice doubted something, he was usually right in his thinking.

"THE COMPANIES ACT, 1948"

Copy.

Special Resolution.

OF

LAMBRETTA CONCESSIONAIRES
LIMITED.

Passed the twentyninth *day of* October 19 55

At an EXTRAORDINARY GENERAL MEETING of the Members of the above-named Company, duly convened pursuant to the provisions of Section 141 (2) of the Companies Act, 1948, and held at 424/426 Kingston Road, Raynes Park, London. S. W. 19.

on the twentyninth day of October 19 55 , the following **Special Resolution** was duly passed :—

"That the Share Capital of the Company be increased from £1,000 to £50,000 by the creation of 49,000 new shares of £1 each ranking pari passu in all respects with the 1,000 existing shares of £1 each in the Capital of the Company; and That the Directors be authorised to allot and dispose of the said new shares to such persons, on such terms and in such manner as they may think fit"

RECEIVED 2-NOV 1955

†WE HEREBY CERTIFY that, to the best of our knowledge and belief, the conditions mentioned in subsection (2) of Section 129 of the Companies Act, 1948, are satisfied at the date of passing of this Resolution, and have been satisfied at all times since the § nineteenth day of December 19 50.

Chairman.
Secretary.

JORDAN & SONS LIMITED
REGISTRATION
2 NOV 1955
AGENTS
CHANCERY LANE, LONDON

Shaw & Sons Ltd., 7, 8 & 9, Fetter Lane, E.C.4
Jordan & Sons, Ltd., 116, Chancery Lane, W.C.2

2 NOV 1955

Part of the agreement with Innocenti was to pay upfront for the orders, and was the downfall of the three men who first started importing it. Even Peter Agg found cash flow difficult because they were beginning to order vast volumes of machines, and it would take quite a while to get the money back into the company. The answer was to issue more shares, and the capital raised this time was huge compared to the first issue in 1950. It was proof that they had hit the big time and were prepared to invest heavily in the Lambretta

A. James Agg
Managing Director

P. James Agg
Sales Director

While the Agg's now had complete control of the company, the team they were building would help make it a success. Peter knew the kind of people he wanted and set about recruiting them.

As more staff were drafted in to cope with the daily running of the company, those at the top had more time to concentrate on building up the empire. Maurice was given the freedom to expand the dealer network and make his own decisions on how he ran it even though Peter was still the boss, and from time to time, there would be reminders of the fact. Derek Miller was still the head of sales for Lambretta Concessionaires, but Maurice played a vital role and had been dealing directly with Peter for some time. Going from an AA man to one of the leading salesmen in a growing company all in a few years was impressive, but how far could he and the rest of them take it in the future?

Chapter Three

The mediator

*A*s Peter Agg, Maurice Knight and the rest of the team weaved their sales magic; the Lambretta became a real threat to the motorcycle industry. The dealer network went nationwide just like they had planned and also that of the service side. Peter Agg had the idea that if you should break down on a Lambretta, you were never more than 30 minutes away from a service agent who could pick you up and get you back on the road in a matter of hours. It was a significant promise to make, especially in more rural areas, but one they were proving was possible. The rest of the industry looked on in envy, most realising they couldn't compete with what Lambretta Concessionaires could offer the customer. Even if they attempted the same business strategy, it would take too much time and investment to implement it; in other words, the Lambretta had a stranglehold on the idea. Towards the end of the decade, it was outselling the most prominent motorcycle brands with over 100,000 units per year, which seemed unimaginable even just a few years back. From the small outfit started in 1950, they had achieved all this in less than a decade and even Peter Agg himself could have never envisaged it would turn out this way. Others looked at their example and wanted to follow suit but not only was it the hard work and determination that had made it happen but the methods they used.

There is no doubt a lot of it was to do with the product itself, and apart from the Vespa, it was light years ahead of the rest of the scooter competition. However, the Lambretta was becoming more than a commuter vehicle; it was becoming a way of life for its owners. Part of this was the club that came hand in hand with owning a Lambretta and an essential piece of the company's structure. It was free to join when you bought a brand new Lambretta from a dealer, and with no subscriptions to pay financially, it didn't cost anything to be a member.

Known initially as Club Lambretta of Great Britain, it was the brainchild of John Cubbon and Francis Gwynne, working alongside Lambretta Concessionaires. Peter Agg liked the idea of an owner's club as it could help sales if it became big enough. Starting in 1953, it proved so popular that Peter Agg felt they should have a club directly linked to the company. Called the British Lambretta Owners Association, or BLOA for short, a membership form was given out with every Lambretta sold.

Derek Guy, who was employed to run BLOA

Initially, the club was outsourced to another company, but due to its success, it was decided it would be run in-house by paid employees of Lambretta Concessionaires. Like the dealer and service agent network, this was another idea to build customer loyalty and had never before done before. Peter Agg saw it as welcoming owners to the Lambretta family and that they would now be part of it. They were not just attending the rallies but allowing them to have an active role in saying how it was run. The man first put in charge of BLOA was Derek Guy, and he would be helped by a committee formed from members of some of the big national clubs. It allowed a more creative input regarding the club and what they thought about the Lambretta. While Peter Agg loved it, Maurice Knight didn't like it one bit. He felt they were constantly interfering with his job, certainly when promoting the Lambretta at the big rallies due to the certain powers they had been given. He just wanted to get on with selling the Lambretta, which he had done successfully over the years and wasn't interested in what he referred to as owners telling him how to do his job.

BLOA was here to stay, and as Maurice sold more machines, it meant the membership grew; it was like a double-edged sword for him. Friction was bound to happen from time to time, and Maurice did vent his displeasure to Peter Agg regarding their interference, but it fell on deaf ears. If there was a BLOA rally happening, one person you could guarantee not to be present was Maurice Knight thinking it wise to stay away and so avoid a confrontation. That aside, Derek Guy happily guided the club in the right direction as it became more prominent by offering incentives to members such as insurance schemes, continental holiday packages and so on. The rallies were the most significant part of the club, with several held annually across the country.

Luton Lambretta club would provide the answer to staffing the BLOA committee

The rallies had two main objectives: to gather members together and showcase what the Lambretta had to offer to the public. It was hoped that this would encourage more sales and become a sort of mobile sales roadshow. The events weren't cheap to put on, and most of the time, the money required was funded directly by Peter Agg out of his own pocket. A significant sum of money well into the thousands, but he saw it as money well spent indeed if it brought in new customers. It was also his way of saying thank you to all those that had bought a Lambretta and got him to where he was now. It also made him feel proud that what he had built up was making so many people happy.

As the club continued to grow, it needed more people on the committee to help run it and those who could dedicate a lot of free time to the cause. Luton Lambretta club was massive and was heavily involved in BLOA and all of its events; logistically, it wasn't too far from the factory either. One of its members was soon recognised as the ideal candidate to help the BLOA cause, his name Pete Meads. He joined the committee in 1960 when the Lambretta was at its peak of popularity, and his enthusiasm was rewarded by being promoted to chairman. It was one of shock; he commented years later, "from being an ordinary club member attending BLOA rallies to being its chairman organising them, I don't quite know how it happened". It didn't matter, as his input would be vital in helping the club grow and the events they held. It was a strange position to be put in because it was only voluntary and not a paid role, but on the other hand, he had many powers and essential decision-making for the club and the company.

Pete had travelled many thousands of miles on his beloved LD and was a regular attendee at BLOA rallies, seen here fooling around at Kidlington

His experience was a vital part and probably one of the reasons they picked him for the role in the first place. His Lambretta career had started several years earlier when he purchased an LD and joined the Luton Lambretta club. They went just about everywhere, much of it long-distance riding, something he relished, allowing him to clock up a staggering 140,000 miles on the LD before trading it in. This devotion to the Lambretta saw him help instigate the overseas rallies that became a significant part of BLOA in the 1960s. With the advent of the series two, the Lambretta had become a genuine touring machine ideal for this type of event, and Pete exploited its potential to the full.

Handing the controls to his wife Sheila for the first time, he seemed pretty calm

No one realised just what an impact BLOA would have when it first started and how it would make Lambretta owners one big family. While some were happy to be members, the majority wanted to be involved in all the offered activities. This was mainly rallies and continental touring, but slowly other activities were introduced, such as endurance and sporting events. Pete was keen to see this side of BLOA grow even further, and actively participating helped even more. The biggest event in the 1950s was the Isle of Man holiday week. Usually held after the motorcycle TT races, the island became a haven for Lambretta and scooter owners alike. Those who wanted to enjoy the spectacle did so, but for the likes of Pete, it was about competing. His thirst for scooter sport grew even more extensive, and he soon became involved in scooter scrambling. Luton Lambretta club had a team with sponsorship and support from a local butcher, giving them the funds to compete. Scooter scrambling was one of the first ways to race competitively against others and was also responsible for the early attempts at modifying and tuning an engine, certainly in Britain.

You name it; Pete could do it, one of the most talented Lambretta riders around. The IOM scooter week had a magnetic attraction, and to competitive riders like him, it was a must

That's one of the reasons it became popular, as those interested in tuning and speed now had the platform to carry out such activities. Pete would compete at as many races as possible and started to be seen as one of the sports role models; It seemed he could turn his hand to almost anything when it came to the Lambretta. If it was good enough for the chairman of BLOA, it was good enough for the ordinary member and showed how BLOA was influencing what Lambretta owners did with their machines.

As the freedom of the 1960s took hold, so did the aspirations of Lambretta Concessionaires, even though Lambretta sales had already peaked. The introduction of the Slimstyle in late 1961 was followed by the TV 175 and then the TV 200, allowing even more outstanding performance. Their influence spilt over into events like regularity trials that were now becoming popular. Usually, run over a 12 or 24-hour format, they were regarded as no more than an average speed test, but competitive riders saw this as a chance to go faster. Pete played a massive part in organising them through Lambretta Concessionaires, usually sponsored by an oil or petrol company as they would benefit most. He enjoyed being part of it, seeing hundreds of riders compete against each other and entering himself when given a chance. For BLOA to continue successfully, it had to evolve and change with the times, and these events became a big part of that.

Number 8 Pete Meads, an expert scooter scrambler

The scooter scene, in general, was beginning to change, and most specifically, it was the younger generation participating. People still wanted a scooter for work or transport, but the world was moving on at a fast pace, and the car was taking the place of two wheels. It was important for those running BLOA to latch on to what their members wanted and ensure they didn't become disconnected from modern-day scooter riders. The mods were doing their bit to put scooters and the Lambretta in the mainstream media, but it would only be short-lived exposure until the press got bored of it. BLOA needed continual support from Lambretta Concessionaires, and Pete did everything he could to ensure that happened. Peter Agg had become slightly removed from it all compared to the glory days as his genuine interest lay elsewhere in motorsport, and his company's success now gave him the funds required to participate. Even so, the Lambretta was still his primary focus business-wise, so he always kept his eye on matters. Even he realised change was necessary for the company to exist the way it had done in the past. He wasn't bored of BLOA, far from it, but it needed to run itself without his input, leaving the committee to get on with the job.

BLOA meetings were held each month at Kingsway Hall, and along with Derek Guy and secretary Jackie Palmer they ran a tight ship

With BLOA still going strong and by 1963 claiming to have 100,000 members, it required excellent leadership skills and dedication to keep things organised; such was the number of people now involved. Not that all members would turn up to an event at the same time that would be impossible to cater for. The rallies had always been well supported and held across the country to reflect the areas of representation. Pete still loved his job, which took up a lot of his time and though he wasn't directly employed by Lambretta Concessionaires or paid, that didn't matter to him one bit. Of course, there were benefits from being chairman and on the committee, but all he wanted to see were owners enjoying what BLOA had to offer. Peter Agg was also grateful for his contribution, which didn't go unnoticed. Even so, he was prepared to make changes not necessarily to the personnel but the club's image if that meant it would grow stronger.

Things were beginning to change; Pete was the owner of a brand-new TV 200, branded the GT 200, getting one the first week they went on sale. This exciting and powerful new Lambretta made directly for the British market was, in his opinion, the best £200 he ever spent and to prove the point later on, even when the SX 200 and GP 200 came out, he refused to buy one as he found the GT to be so good. No doubt this new model was a shot in the arm for BLOA, as many new members joined when they purchased one. It also meant the sporting events would benefit from its introduction as well as reliability trials. The trials became like a race meeting; such was its power, and riders wanted to exploit the engine's potential. At the same time, not forgetting this was now the ultimate touring Lambretta meaning overseas events would become even more popular.

Running the reception for the BLOA Southend rally 1963 with his brand new TV 200 proudly on display

In charge of the reliability trial at Snetterton 1964, Pete donned a deerstalker hat. As he described," not only did it make me stand out in case someone needed my attention, but it also added a bit of the old-time racing feel about it"

At the time, Derek Guy was still the man in charge of running the BLOA, and nothing happened without his final say, even though the rest of the committee helped in the decision-making. Remembering any wrong moves within the club would fall on his head as he was being paid to get it right. Meetings were held at Kingsway Hall in London every month, where members were invited to openly discuss how the club was run or could be improved. Pete would run the meetings and report back to Derek with his findings and whether or not they should make any changes. Pete was highly regarded because he could be relied upon, becoming the middleman between the members and the owners, a mediator.

By now, Derek was in high demand, and his success at Lambretta Concessionaires hadn't gone unnoticed by others. Before long, he was being headhunted, and the company that sought after his skills was way too big to turn down, namely Castrol. Seeing it as a new chapter in his life and an opportunity he would regret if he didn't take it on, he accepted their offer. Though sad to see him leave, Peter Agg realised it would be difficult to keep hold of Derek and didn't want to get in the way of him furthering his career. Regardless, plenty of people would jump the chance of joining a successful company like Lambretta Concessionaires, so finding a replacement may not be too difficult. However, the question remained: what would that mean for BLOA and its future?

In the short term, nothing, and it carried on the same as it always had; with a strong committee in place, it was in safe hands. However, it was too big to leave with unpaid volunteers as they couldn't carry out the required day-to-day administration. The solution seemed simple enough to get someone else within Lambretta Concessionaires to take over the reins, similar to when Derek first took charge. Peter Agg already had someone in mind who this person would be, someone who had played an essential part in the company over the past several years. However, it was an outsider, not someone already employed by the company, which surprised many. Not only would this person make an everlasting impact on BLOA but Lambretta Concessionaires as well.

Standing with Ken Peters on the way to the Eindhoven rally in 1963. By now, the club had a new person in charge, and the leg shield banners indicated a change to the club's name. Pete being the gentleman he was, got on with his job; all that mattered to him was the club continuing to be a success

It was a bittersweet moment for Pete Meads as perhaps he should have been offered the job, but it never materialised. Even if it had done, it would have meant relocating, and he already had a job and comfortable lifestyle near Luton; perhaps the risk would have been too significant to take on. For him, that didn't matter, and he welcomed his new boss with open arms as far as he was concerned; these were exciting times ahead. Not that he was disrespecting all the hard work Derek Guy had done, far from it, but a new person running BLOA with fresh ideas could only be a good thing. Changes would be a foot everyone was well aware of that, a new person with a different perspective but perhaps not how Pete or anyone else involved could have ever imagined.

Despite having moved on from his role at BLOA, Pete always stayed loyal to the Lambretta cause as well as his TV 200, seen here competing on it at the Isle of Man in 1970

Receiving his winner's trophy at the IOM 1970

Regardless of who that person would be, it was Pete Meads contribution as an unpaid volunteer that had kept the BLOA success story going for the past few years. He wasn't the only one; many enthusiastic helpers played their part and should never be forgotten for what they did. However, what he achieved and its significant impact on the Lambretta story will always be remembered.

Chapter Four

The advertiser

*R*obert Kitson Wilkinson Left school in 1948 and started his working life as an office boy at solicitors located in Pall Mall, London. It was just a stopgap as he wanted to go down the career path of joining the army, following in his father's footsteps. His goal was to become part of the military police, and in 1951 having been called up for national service, it seemed like it would become a reality. However, he was dealt a devastating blow during the medical as they discovered he had a heart murmur and so was rejected. He was in total shock, and it would take a long time to get over; more importantly, it meant a new career path needed to be chosen. Getting used to the daily rituals of working life would help, and he was no fool, a realist quickly working out that you have to start at the bottom rung of the ladder and work your way up to the top. Before too long, he had changed companies and, this time began employment in an advertising agency; while working here, he realised where his destiny lay.

Advertising played a considerable part in any business, not so much through television as that was still in its infancy but in the press, where it was of paramount importance. Bob not only found it exciting but realised he might have a natural talent for being creative, an essential part of the job. That was a few years off, but, in the meantime, he worked his way up the ranks making it to assistant production manager, and this is where he began to learn the ropes, gaining valuable experience in a fast-paced industry. The progression saw him work for several firms before finally landing a role at Smee's in Manchester square. It wasn't about leaving one company after another and trying to move on to something better, as that is the wrong way to progress. This time it was a case of being headhunted as his talents hadn't gone unnoticed within the industry. Smee's were a big player in media advertising, and Bob landed the role of assistant account executive, which was a big step up. He would work under the leadership of Desmond Smee, who put him in charge of one particular account under the name Lambretta.

Des Smee was a good friend of Peter Agg, who happily pushed the Lambretta advertising their way, spending a lot of money with the company in the process. It was 1958 when Bob took the account over just as the Lambretta became a dominant force in Britain. With the advertising budget for Lambretta products being hugely increased, it would mean a great deal of work for him to get the adverts done on time and make them stand out in a competitive climate. By now, the Lambretta had blanket coverage across all national and regional newspapers, not forgetting billboards, sides of buses, and train stations, almost anywhere it could be seen. It was about constantly reminding the general public why they were better off if they bought a Lambretta.

Whether it was saving time on a journey, having an economical vehicle or the freedom to go where you wanted. The message was pushed across almost weekly, not forgetting the offers on cheap insurance and accessories to improve the riding experience or anything to do with the Lambretta lifestyle. All of this required new slogans and angles to get the point over, and the person in charge needed a creative mind to keep producing adverts that made a significant impact. Bob realised how much revenue the account bought into the company just by the sheer volume of work and knew he had to deliver. While it might have been a high-pressure job, if he got it right, it could lead to much greater things in the future and having already been headhunted, his stock value would only go up.

Bob didn't own a Lambretta, nor had he ever ridden one, but that didn't matter; what did was that he portrayed its image in the right way. However, he started to feel hypocritical about the situation because he was producing adverts aimed at the masses but had nothing to do with the product itself. The only solution was to acquire one, which Des Smee quickly sorted through his contacts with Peter Agg, allowing Bob to get one at a heavily discounted price. From that day on, he would make the 14-mile round journey to work and back on his trusty new steed come rain or shine. An LD 125, he loved its smoothness and effortless performance through the busy London traffic and could see why it was selling so well.

For the next two years, he cleverly produced a long succession of adverts and slogans, which won the public over time and time again. It seemed like the advertising world couldn't do without him and certainly for Smee's and Lambretta Concessionaires, so it was a surprise when he not only left but got out of the industry altogether. Having got married when he first joined the company in 1958, sadly, two years later, the marriage had broken down, and Bob felt he needed to get away from it all, London included. He took time out to move to Ross-on-Wye becoming a bar manager, a far cry from his rapidly progressing career. A wage was the most important thing; so be it whatever type of employment gave him that.

A draft for the Lambrettability campaign one of the many ideas Bob came up with

He may have been recharging his batteries, but it wasn't long before a knock at the door tempted him to return to the industry he had turned his back on. This time it would be as an account executive and not with Smee's but with one of their ex-employees who had set up his own agency, Lawler & Borridail. It was too good an offer to turn down; gone were the days of being an assistant; now, he was in charge of a department. It was 1962 and the decade where it was all happening was primed and ready to take off with Bob sitting in the middle. He soon got to grips with the job, but before he could get his chair warm again, there were those circling who wanted to hire his services. This time it was slightly different as Bob was approached by a man called Phillip Keeler, who worked in the public relations department for a huge company going by the name of Lambretta Concessionaires, whom Bob was familiar with during his time at Smees's. They sent over a smartly dressed young man by the name of Peter Baker, who was in charge of the accessory department, to conduct negotiations as he had seen Bob's true potential first-hand.

Bob being introduced to and shaking hands with Pete Meads after taking over the running of BLOA in early 1963.

When he worked on the Lambretta advertising account, Peter Baker had leased with Bob. Accessories had become a considerable part of Lambretta Concessionaires business, and with so many products being made, they were highly lucrative for the company. Together with Bob, the two men produced many brochures and adverts for the accessories they sold. Bob recalled, "It was difficult to list and categorise them all, so Peter would visit my office regularly to find the best way to showcase what was being offered". At the time, it was just getting on with the job, but there is no doubt Peter Baker would considerably impact Bob's career both now and in later life.

Phillip Keeler had been aware of Bob's talents for some time and remembered how good he was at producing their designs, the reason they approached him. However, the job they initially offered him had nothing to do with advertising; instead, he was asked to take charge of running BLOA. Lambretta Concessionaires, at the time, were going through many management changes, not because there was anything wrong with how the company was run, far from it. They were so successful that those at the top were regarded as business gurus and were being headhunted. Derek Guy ran BLOA, but he was on his way out, having been given the job of running Castrol, which was too great an opportunity to turn down. With the BLOA position now vacant, Peter Agg thought it wise to bring in someone from outside the company. A young person full of energy and dynamism who could take a different approach to how it was run. Nothing was wrong with BLOA, just that a new person could inject fresh ideas to improve the club in specific ways.

It didn't take much persuading him either, as Bob duly accepted the offer of becoming an employee of Lambretta Concessionaires in the early months of 1963. He was inheriting a colossal club with one of the most extensive memberships of any single make machine in the country and wasted no time implementing new ideas. The way they ran it was fine, and the organisation was at the highest level that didn't need changing. What did was what it offered in terms of its appeal and its branding, as he felt it was looking a bit old-fashioned. BLOA had been in exitance for a decade in one form or another; what fitted well during the 1950s was becoming dated ten years on. In Bob's own words, "it was all a bit mundane with egg and spoon races and pleasant gatherings; it needed spicing up".

Bob didn't have a Lambretta to start with as he had sold the one Des Smee sorted for him, but Peter Agg soon solved that issue as the head of BLOA couldn't be seen not to ride one; what kind of message would that send out? There was a pool of Lambretta's used by the company for advertising and promotion, and Bob was given a choice to use whatever one was available at the time. Before anyone else, he would get his hands on anything just launched, such as the TV 200, a perk of the job and a good one at that. Bob used the first few rallies and events to take stock of what he had inherited and, having met up with Guido Candello, the head of Lambretta Club D' Italia, realised what changes needed to be made to make improvements. The LCI were far more modern in running things, and Guido helped him implement them into BLOA. Bob's first and most significant change was to the name, which he felt didn't reflect what it was all about. British Lambretta Owners Association was acceptable for the home-grown audience, but now, the worldwide Lambretta story had become a truly international affair and the name needed to reflect that.

Bob soon got to work, and once he had a Lambretta to use started to promote BLOA; however, that would quickly change seen here at Biggin Hill and for the first time displaying a leg shield banner with Lambretta Great Britain written on it

Bob second from the left and proudly displaying the new logo and the one that is still used to this day

Working closely with the LCI inspired Bob to make changes to the LCGB, modernising it and making it more appealing

Bob next to Rex Collier, the Power and Pedal Magazine editor at the Ostend rally. Working with these types of people would help promote the club in mainstream magazines

At the same rally with Guido Candello (sitting down third from the left), who would work closely with Bob in making the changes he wanted to with the club

He changed it to Lambretta Club Great Britain to bring it in line with other nation's clubs. Mike Karslake, who had played an active role from the start through its first incarnation until it became officially known as BLOA, didn't want the change to happen and made his point known. While Mike was an important asset to BLOA, he had his own company and wasn't employed directly by Lambretta Concessionaires. The structure of BLOA wasn't what it seemed either, as a member didn't have to pay to join. It was free membership for anyone purchasing a new Lambretta. Many aspects of it were run by enthusiastic members such as Mike, so while he helped organise events, he didn't decide how it was run. That was down to Bob, he was the man in control now, and he had to put his stamp of authority on it as it would be seen as a weakness if he backed down. Mike wasn't angry as they both wanted what was for the best for the club; he was finding it hard to accept that things had to move forward and wondering how the members might react to the change. The answer soon came as the general agreement was that it sounded right and, with a new logo design, looked more modern. Mike accepted it was the right choice to make and soon put his early rumblings to one side as he endorsed the LCGB with open arms.

Members rallies still had huge participation but getting everyone's attention was a problem; Bob rigged up a mobile PA system so he could be heard at all times

Such was his generosity when riders were travelling over from Europe; he would let them stay at his house even if they were from the other camp (Vespa owners)

One of the main reasons for the change was the international rallies happening more frequently and something Bob wanted to take much further. The name Lambretta Club Great Britain fitted perfectly into how he wanted to sell the image abroad. With rally attendances in the thousands, it now felt like Britain had an accurate Lambretta representation. It would also help when being the host country to make the rallies more appealing to overseas riders. Everyone knew the landscape regarding transport had shifted, and it was felt that the Lambretta image needed to move with the times; to survive on sentiment alone wasn't enough.

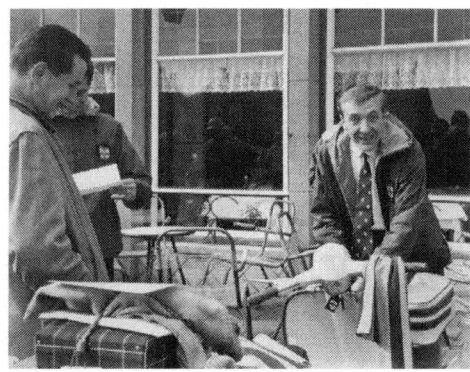

Bob enjoyed touring abroad and learned from the international rallies how to make those held in Britain even better

The plan for member's rallies was to modernise them by way of what went on and include a more significant sporting element. With the introduction of the TV 175 and TV 200, Lambretta owners now had more powerful machines than ever before, so why not capitalise on the fact? There was the option to increase the regularity trials and navigational events over long distances, all of which encouraged more members to enter. The new breed of Lambretta was more than capable of taking on such feats, and it seemed owners wanted to join even if their machine was brand new. Bob could see that members were changing how they used their machines and adapted to them rather than leaving the club in the past.

As 1963 rolled on, it became evident that Bob's introduction had a significant effect and injected a new lease of life into the club. He was creating an aura surrounding the Lambretta, which hadn't gone unnoticed by others within the company. The GT 200 (TV 200) had been a risk and was only implemented because of the pressure Peter Agg had put on Innocenti to produce it. Built for the British market only if sales failed to take off, then Innocenti could quickly stop production without too much damage to their brand. While Lambretta Concessionaires were confident not only in the GT 200 itself but also in their advertising, they were convinced it would be a success. This is where Lambretta Club Great Britain helped because if they could be seen to endorse it, there was no better advert.

Pete Meads had bought one when they first came out and was still playing an active role with the rebranded club working almost like Bob's righthand man. His GT 200 appeared almost everywhere at rallies, trials, you name it, even used by Bob when he got married as the chosen choice of transport to the church. Peter Agg never doubted the importance of the club, seeing it as the connection from the company to its customers and liked the way it was being used regarding the GT 200, which was, in effect, his baby. Bob had taken on almost two roles, running the club and promoting the Lambretta, and the latter was making waves throughout the company.

When Bob married Ursula, they used the TV 200 belonging to Pete Meads as their wedding carriage. Pete also baked their wedding cake as another kind gesture

The TV 200 appeared everywhere here, being used by Guido Candello when a guest at the Southend rally, Bob and Pete standing either side of him

When it seemed like all the management changes had taken place, Dick Hullett now handed in his notice to quit. That meant Lambretta Concessionaires needed a new advertising manager, and Bob was seen by many as the natural replacement. The experience he gained working on the Lambretta account at Smee's back in the 1950s should have been enough to get him the job on his own merit. More than that, they had experienced him working his magic on transforming BLOA into the LCGB and seeing how he worked so meticulously.

There had been problems getting the right person for the job in the past, and at one point, a man called Martin Attlee had been given the job. He was the son of former prime minister Clement Attlee, and it was rumoured Peter Agg had taken him on more for his political connections than his advertising skills. He struggled with the role, eventually leaving, and it seemed to be a vacancy that was difficult to fill. Dick Hullet had done an excellent job, but with him giving in his notice, they needed to find the right person this time, certainly with the industry's changing landscape.

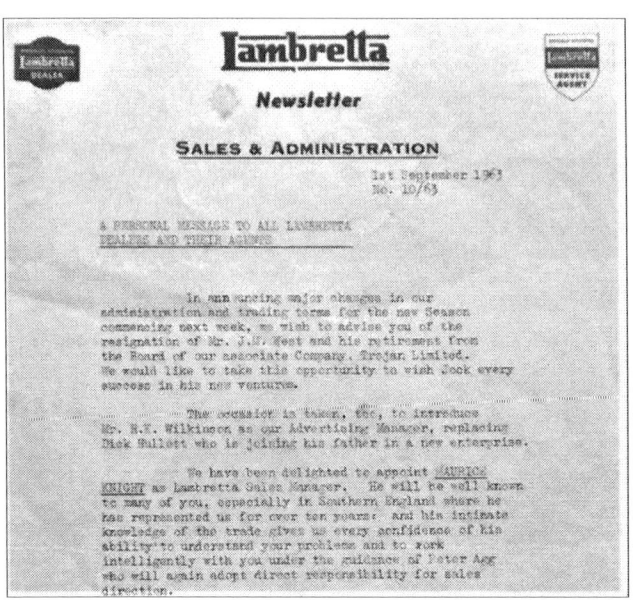

The announcement that Bob was taking over the role of advertising manager was relatively short and sweet. Below that was the news that Maurice Knight had become sales manager simultaneously. They would form a great partnership over the coming years and play a vital role in taking the company forward

That said, Bob was officially announced as advertising manager in September 1963 and was eager to get his teeth into the new role, which he was very proud to accept. He would still run the LCGB, so he would have two jobs within the company, which was a lot to take on, but he was sure he could succeed with both. Though these were good times, he still had to answer to Peter Agg but was given free rein in how he ran each department.

Peter Agg didn't see the point in interfering with Bob's decisions when he took over BLOA, and it would be the same with the advertising department as long as nothing was too draconian. Even so, he would strictly monitor everything to ensure he wasn't overstepping the mark, just like he did with all the others who ran departments within the company. By this time, there was a lot of competition beginning to hot up in the two-wheeled market, not forgetting the Mini motorcar had taken off and was making four-wheels affordable to the masses. The two main threats were Vespa, whom Peter Agg referred to as "our vulnerable friends in the west", and Honda. The Vespa had been the thorn in the side of the Lambretta for years, and they were accustomed to their advertising tricks and knew how to handle them. The concern from Honda was the unknown quantity, as little was known about the Japanese company and its products.

They had recently introduced the "Honda Cub" into the British market, and though it had bigger wheels, it was seen as competition to the Lambretta. It worried Peter Agg the most because he didn't know how serious the company was about trying to take a significant share of the market. The public didn't know much about Honda or any other Japanese two-wheeled manufacturer either, so in the meantime, they weren't regarded as a threat, but that didn't mean things couldn't change in the future. A lot depended on what Innocenti did as it was their products Lambretta Concessionaires were selling. Peter Agg may have influenced them into building the TV 200, but he couldn't make wholesale changes regarding new models. That was down to Innocenti themselves to keep investing in and developing the Lambretta.

Bob took no time settling into his new role as advertising manager, and one of his first campaigns was "let yourself go on a Lambretta". While models from local agencies would be used for the photoshoots, Bob himself often joined in, like here

Two images of Bob leading the British contingent at the Rome rally in 1965. He loved all the rallies and the international ones more so. The problem was they took up a lot of his time, and he had to juggle the two roles he now had within the company, but it couldn't stay that way for too long, such was his ever-increasing workload

"One week, it was organising an LCGB rally the next, it was promoting the Lambretta"

Presenting Vespa Club of Britain boss Ian Kirkpatrick with a trophy labelled "burying the hatchet". The idea was to show there was no animosity between the two clubs. With any exposure in the scootering press, as usual, Bob saw it as a great PR opportunity

For Bob, it was a case of maximising adverts' impact and spending the budget wisely. The days of blanket coverage were long gone, and the money had been slowly reduced each year, making it much harder to get the point across to the buying public. It would be the mainstream media they would concentrate on to get the biggest possible audience with the least amount of money spent. There was no point in advertising on the side of a bus where only a few people would have exposure; national newspapers and magazines were where it needed to happen to target a vast audience in one go. Putting a direct advert into national papers would take a considerable amount of the year's budget with very little left over for anything else. So, to make it work required free advertising by way of stories that the press wanted to write about. Any article could have a small advert placed next to it to get the maximum effect; that wasn't the issue; the stories themselves were. Bob needed to find owners doing something different with the Lambretta and create a story based on them. He was good at thinking up adverts and slogans for the Lambretta, but now, he needed others to help him make a significant impact. Peter Agg was sure he had the right person in charge of doing so but could Bob deliver what he wanted and successfully?

Chapter Five

The professor

*B*y the mid-1960s, Lambretta Concessionaires success meant it was employing more than 1200 people at its Croydon factory. This figure was divided across the Trojan Group of companies, but the Lambretta was the backbone of revenue created even though it wasn't bringing in as much as it did in the boom years of the late 50s. Thanks to the recent success of the TV 200, it had given a welcome boost to sales and opened up a new market in scooter ownership. Having a 70mph Lambretta or so the company claimed, was a big lure to those wanting performance. With Arthur Francis having just released a performance version called the "S Type", things were about to heat up even further.

What it now needed was for the company to take advantage of the TV 200 in any way it could, and while that would happen through advertising, it required a big budget if going down that route. What if the owners could do things with the TV 200 indeed that would create media interest and, in effect, be a form of free advertising? Don Noys had done that when in 1964/65, he broke several world speed records on a self-tuned version, creating tremendous interest. It was probably his exploits that led to the Filtrate/Lambretta Concessionaires Atlanta V project taking place not long afterwards. Though that was doomed to failure due to lack of organisation and proper testing, using a female pilot, Marlene Parker, had created much publicity for the company. The Daily Express had come on board, followed by several motorcycle magazines picking up the story, not so much the attempt itself but the selection process generating a big media frenzy. The Daily Express carried several articles on the story throughout 1965, which was priceless for the free advertising it gave Lambretta Concessionaires.

Two years before these daring exploits, a young man named Anthony Tessier was venturing increasingly into Lambretta ownership. Secretary of the Greenford Lambretta club in the London borough of Ealing, he was located in the right place to become inducted into the world of BLOA. As the Greenford club was large and active on the national scene, Tony was invited to join the BLOA national management committee around the same time that it was being rebranded and renamed Lambretta Club Great Britain. That's when he became noticed by Bob Wilkinson, who was in the process of putting a team together for the epic Milan Taranto endurance trial. The prestigious event was highly regarded in Italy, and for the 1965 event, Innocenti and the Lambretta club d'Italia wanted a British team involved. Not only would this put the LCGB in the limelight but also Lambretta Concessionaires, and so, keen to get maximum media exposure Peter Agg approved and funded it through the company.

There would be nine riders in the British contingent, and Tony was asked if he wanted to be one of them. He jumped at the chance and was proud to be part of a team representing his country, but there was a problem, he had crashed his machine at Snetterton a couple of weeks before. Bob told him not to worry as there was a demo TV 200 at the factory he could use and that he could have it before the event to check it over and get used to it. Not that it mattered to Tony; getting free use of such a Lambretta was an honour.

Bob arranged shipping of the machines and riders by air so they would be refreshed and ready to go before the start of the race. He also had the job of following them in the company's Commer van, and as he put it, "I completed the Milan Taranto in 1965 but unfortunately on four wheels instead of two". Expressing it was a gruelling journey couped up in a little tin box, and not sure if it was more tiring for him or the Lambretta riders. They needed backup in the case of a breakdown, which shows the considerable effort Bob and others made in promoting the club and Lambretta brand.

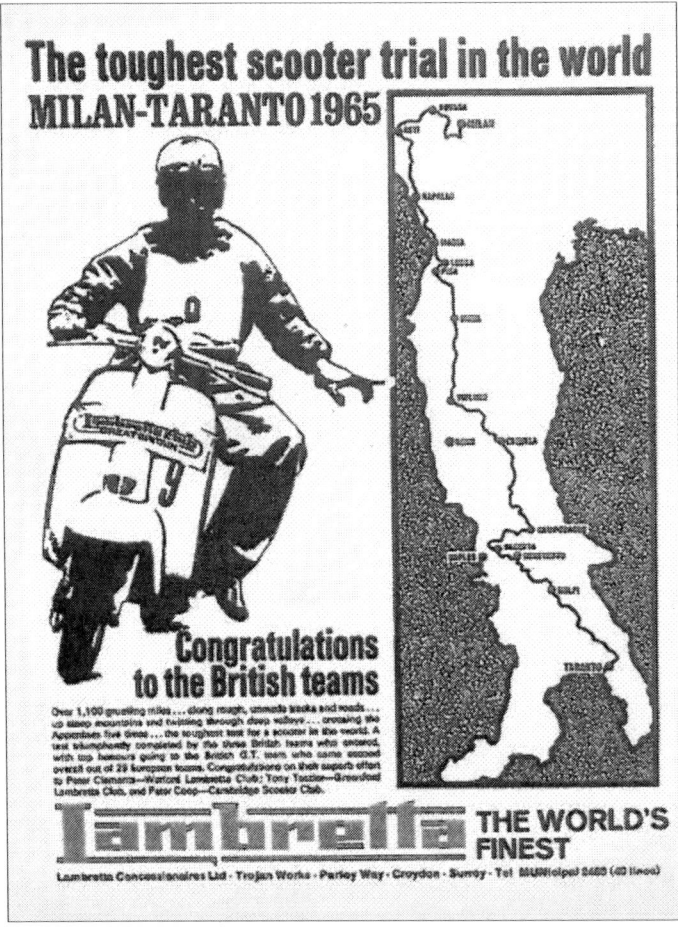

The Milan Taranto trial was held in great esteem in its home country, but Innocenti wanted a team from Britain to compete. While funding for the team would come direct from Lambretta Concessionaires, they made sure they got their money's worth by using the event in advertising, focusing on how durable the Lambretta was

All nine machines used were lined up for a photo shoot at Southend airport before being loaded onto the plane

Preparing for flight. With great care, each machine was loaded on board before being flown out to Italy

Tony soon realised the importance and intensity of what he was about to embark on. The Milan Taranto trial was a prestigious event, and there was tremendous competition, especially with the Italians who wanted to beat the British team on their soil. There were things to overcome before they started, and he explained, " It didn't help that all the maps and road books were Italian which was necessary for us to navigate correctly (and none of us spoke Italian)". Tony was an astute person who just got on with it like the rest of the team and was eager to repay Bob's faith in him. In the end, they came second overall to Lambretta club Cagliari (Sardinia) and only by a fifteen-second margin which was nothing over the 1000 km distance of the trial.

Team members (3 teams of 3) waiting before a time control en route to Alberobello, Puglia

The idea was to return in 1966, but sadly, it never materialised as the funding was not forthcoming from Innocenti, who backed it. For Tony, it was back to earth with a

bump as the TV 200 immediately had to go back to the factory; Ken Peters was using it for an attempt at the London to Milan in record time challenge the following week. Not that he minded, as he was still on a high from his escapades, and all the press coverage had put him and the rest of the team on a pedestal. They were almost like superstars admired by the rest of the club members, and why not? They deserved it.

Bob Wilkinson travelled with the team all the way and checked the timing sheets to see how they were performing after each stage

Tony's team with Bob Wilkinson at Geneva airport before travelling back to Britain, left to right Tony Tessier, Bob Wilkinson, Peter Clements, and Peter Coop. The team finished second in the event, with Tony 13th overall

His Lambretta journey wasn't over as he became part of the workforce at Lambretta Concessionaires thanks to Bob, who got him a job there. By now, it was the summer, and they needed workers in the store's department as things were getting hectic. It wasn't the glamour of a high-level job within management; quite the opposite, shifting crates of spares around and unloading trucks. It meant money was coming and being involved in the Lambretta story, but a far cry from the glory of the Milan Taranto experience, and Tony didn't like it one bit. At the same time, he had met up with another employee who worked there, Malcolm Clarkson and they soon became good friends. Malcolm had other ideas and left not long after to set up his own Lambretta-based shop, namely Supertune. Within weeks the business began to make a big impression on Lambretta owners due to their performance tuning products and race-liveried machines. Tony struggled on at the factory, but one day he was asked to unload a large quantity of series three U-bends on his own. It was the final straw in the heat of the summer, and he was exhausted, so he quit.

There was no animosity towards Bob, and he was grateful that he had given him the opportunity, but there was no way he could carry on with what he was doing. Sometimes that's what happens in a big company as Lambretta Concessionaires had become. Bob fully understood, and they remained good friends, but for now, Tony would go down a new path even though they would meet up again in the future. Remembering the friendship he had built up with Malcolm Clarkson, he visited him, and Tony was employed at Supertune as a mechanic before long. It wasn't all it was made out to be with a chaotic workshop and environment, along with Malcolm, whose financial side of running the business wasn't run very smoothly. Tony soldered on for two years but realised he wasn't going anywhere with this job, so once again, he moved on. Now out of the scooter trade altogether, it seemed that was that, but it wasn't long before his old friend Bob Wilkinson was back in contact not with another job offer but to do another trial in Italy. This time it was the Moto Giro d'Italia which was as highly regarded as the Milan Taranto. It had ceased being an all-out road race in 1957 after the Italian government had banned such events following the Mille Miglia tragedy, where several people were killed.

Back on the road with Bob once again. Loading the trailer in the workshop at Lambretta Concessionaires in readiness for the long journey to Italy. Left to right John Forster, Paul Chambers, Colin Tomlin, Bev Flanagan, Nev Frost with Tony behind the camera lens

The event had returned in 1966 as a six-day regularity trial, and in 1968, the LCGB were intent on returning to Italy to try and win and beat the Italians. This time, it was a smaller contingent with just six riders, but once again, they would be backed up by Bob Wilkinson and the trusty Commer van. Tony had his own Lambretta, this time an SX 150 which was heavily modified and perfect for the endurance race he was about to take on. Despite their best efforts, they failed to win, but all who took part agreed it had been a great experience. Sadly, it was the last time Lambretta Concessionaires would fund a team to go to Italy as there were more demanding needs back in Britain; however, this wasn't the last time Tony would have a connection with the company.

At the hotel in Chianciano Terme before the start of the event and once again, the trusty Commer van piloted by Bob would follow behind the riders as backup

By now, he was a student handicraft teacher at Shoreditch College and had been working for some time on an idea about hovercraft design. It was part of a thesis he was doing in collaboration with a friend, but the Lambretta connection was about to happen once again. To drive the huge fans that created the lift and thrust for the hovercraft required a powerful engine, and Tony, an ardent Lambretta fan, wondered if he could integrate the engine from one into his project. It required a lot of thinking and planning to see if it was possible, but he soon realised to create enough power would require not one but two 200cc Lambretta engines working in unison and so be quite expensive. They would also require modifying, not tuned for more power but removing the transmission with the idea to drive the fans off the front sprocket. All he now needed was the funds to purchase the engines.

They would be pretty expensive to get hold of, and Tony's limited budget didn't stretch that far. So, once again, in stepped Bob Wilkinson, who saw a vast PR opportunity taking shape, and with Peter Agg's nod, they helped fund the project. It was just the kind of publicity they needed remembering back to the appeal for owners to come up with newsworthy items, and a Lambretta-powered hovercraft was as wacky as they came. So not only would the factory supply two engines for the project, but they would also paint its body, and that is when Tony conceived the "Lambrettair" idea. Painted along its side in the same logo used on the Grand Prix, it stood out as a great advert, and with the Grand Prix having just come out, it was the perfect time to get free press coverage.

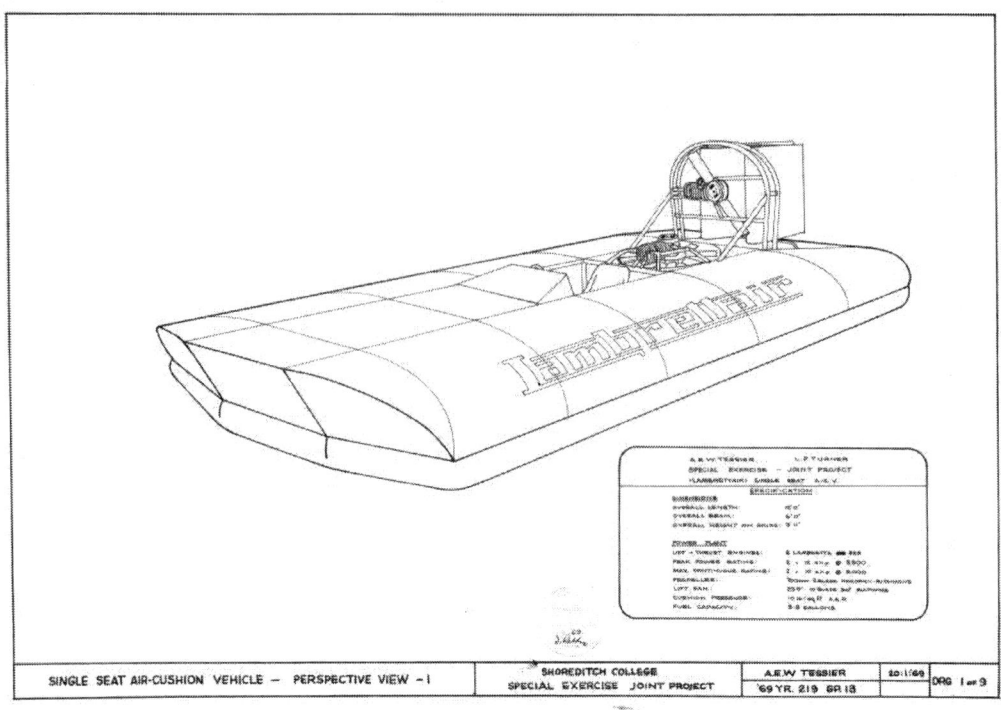

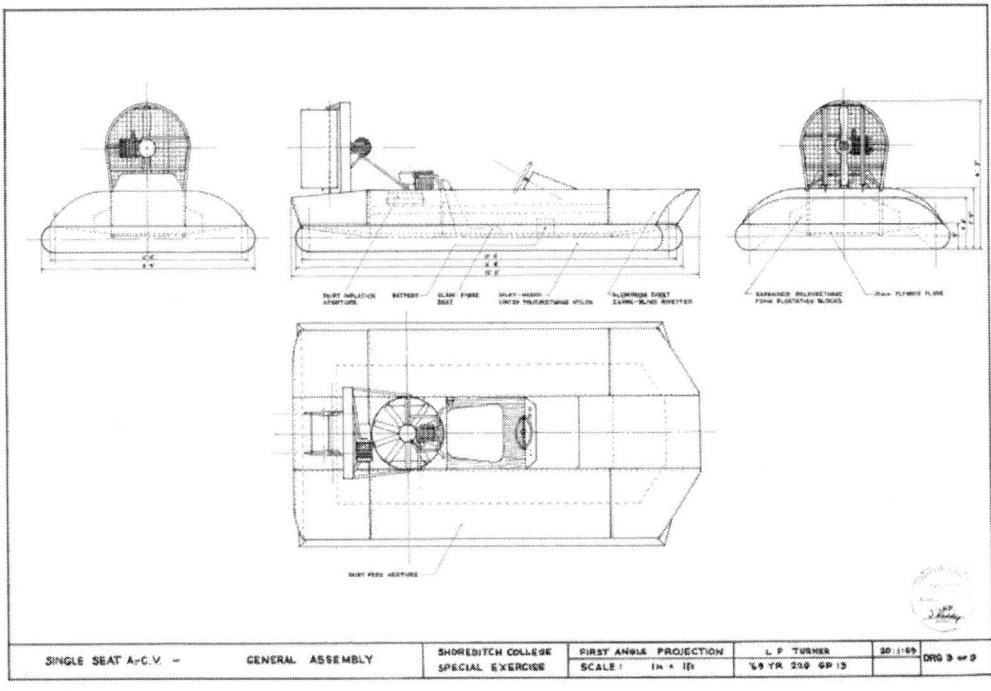

Tony's drawings for the hovercraft powered by two modified TV 200 engines, and named by him "Lambrettair". At school, he was known by his classmates as Professor Tessier because he was always thinking up ideas and inventions now, in later life, that had become reality

Bob wrote a press release showcasing Tony's invention and highlighted that it was powered by two 200cc engines, the same as used in the Grand Prix. According to Tony, that statement wasn't exactly true as they were using a couple of unused TV 200 casings that had been lying around in the factory, but it didn't matter too much; what did is that the hovercraft worked. Those at the college didn't think the design would create enough lift, but little did they know how well it had been thought out. The day of reckoning would happen when the examiners tested it to see if it could carry out the task it was designed for.

Lambretta

Lambretta Concessionaires Ltd
Lambretta House Purley Way Croydon
CR9 4HD England Phone:01 686 2499
Telex 21138 Cables: Lambretta Croydon

SCOOTER POWERED HOVERCRAFT TO MAKE TRIAL
TAKE OFF AT BRANDS HATCH ON APRIL 3
Press Preview for International Motor Cycle Show

A scooter without wheels which rides on air over land or water is to make its debut at Brands Hatch circuit on Thursday, April 3, at the Press preview for the International Cycle and Motor Cycle Show, opening at the Hotel Metropole, Brighton on April 5.

Known as the "Lambrettair," it is powered by two 200 c.c. Lambretta engines - the same engines which power the new Lambretta Grand Prix scooter which is also being shown for the first time at the preview.

It is being built by 25-year old Tony Tessier of Greenford, Middlesex and 21-year old Laurie Turner of Leek, Staffs, both of whom are student handicraft teachers at the Shoreditch College of Education, near Windsor.

Tony got the idea three years ago after preparing a thesis on light hovercraft and he and Laurie started designing "Lambrettair" in September 1967. They spent 12 months on the drawing board, during which time the craft they were planning to build was also the subject of a symposium at the Hoverclub of Great Britain.

As both these students are on a three-year course in advanced metal-work, they decided to construct their hovercraft in metal - most home-made ones are of wood - but made sure, first of all, of obtaining the advice and re-assurance of a metallurgist from the British Hovercraft Corporation.

They started building in February this year and have spent 500 man hours on the job. Their craft is made of 250-ft of 20-guage steel tubing and 34-sq. yrds. of 26-guage aluminium and is 12-ft long, 6-ft. wide and 1-ft. 3-in. high. It weighs 420-lbs.

Bob made sure he gained total media exposure with the Lambretta hovercraft idea

Tony at the controls of Lambrettair on Marine Parade, Brighton, a demo for the International Motorcycle Show

The engines were heavily modified, and the gearbox and transmission were removed, being powered from the front sprocket. A pull cord mechanism similar to that on a lawnmower was adapted to start them. The original exhausts were removed with ones from a motorcycle fitted, and though they were baffled, the noise both engines running at the same time created was ear-splitting. Note the fuel tank precariously sitting on the top, which was taken from a J range Lambretta

In the paddock at the Brands Hatch international motorcycle show press day. Peter Davis, who was now the LCGB Secretary in the pressure suit with Tony next to him, and at the controls Laurie Turner, Tony's co-designer, and builder of the project

Peter Davis is at the controls, with Bev Flannigan as a passenger. Bev had become a huge scootering personality, and any chance Bob Wilkinson had of including her in PR events, he would do so

The test would take place at the local football field, and it had to carry two people to pass. One of them said, "that will never lift two people", and Tony, somewhat annoyed at the comment, replied, "oh really"? "Let's see what will happen if we put six people in it then". So, in they got, he fired up the lift engine, which rose immediately to the full hover height clearance. Increasing the speed of the thrust motor, it moved forward, albeit slowly at first, with looks of panic from the examiners followed by "okay, we believe you!"

Now it was proven Bob stepped in to maximise its potential regarding publicity for the Lambretta, so a press shoot was planned for early April 1969. Though a man hadn't yet landed on the moon, the space race was the predominant theme that year, so the idea was to have the driver wear a space suit-like design to make more impact. The venue would be brands hatch, and with a considerable crowd gathered, the hovercraft proved once again how well its Lambretta-powered fan allowed it to glide along. The only downside was the enormous dust clouds the fan threw into the air and the problems it caused. Initially, the air filters had been left off, and Vega ones fitted instead, but they choked the engine, so it was decided to leave them open-mouthed. The problem was that the dust got sucked into the engine and quickly wore out the barrel and piston; thankfully, spares were on hand from the factory free of charge.

Tony proved his idea worked, and he and Bob got to work together again. Though he was only briefly employed at the factory in the middle of the decade, his exploits played an essential part in promoting the brand. It validated Bob's idea that those outside the company could be equally as important to the cause through what they did. Many people were of a similar ilk, but Tony's involvement was over a much extended period, and in many different ways, his contribution to the Lambretta story was hugely significant.

Chapter Six

The dream team

*T*he early promise of the 1960s continued to gather momentum; if anything, mid-way through the decade, it was getting even more substantial. Everything changed so quickly; what was headline news and ground-breaking one week was forgotten about the next as something just as vibrant came along. Music and fashion led the way with everything else being sucked in behind, and for those involved, they were exciting times. It required inventive and creative people who could constantly evolve and come up with new ideas, those who couldn't simply got left behind.

For Lambretta Concessionaires, the rules were exactly the same; there was no exemption because they could no longer rely on the Lambretta alone to sell itself. It was now becoming a battle to maintain the figures, and with Britain being the second biggest market, Innocenti was heavily dependent on them. It heaped a certain amount of pressure on the company but more so on Peter Agg, who had built up an impressive empire and wanted it to remain. The Trojan Group had an extensive portfolio of brands, but there was no doubt the Lambretta had been the commodity that helped to create it all. With other parts of the business doing well, they were no longer solely reliant on the Lambretta to generate profit even though it was still the mainstay. Peter Agg hadn't built up the company to this point without being an astute businessman but realised selling the Lambretta was far more complex than in the past. He relied on those he employed to do the job and make it happen; it seemed he had the right people in place. The two he put his faith in the most were Maurice Knight and Bob Wilkinson, and they were expected to deliver.

Bob would gain maximum exposure for the Lambretta brand wherever possible. The Atlanta V project was Filtrate oils idea with Lambretta Concessionaires helping out. Using a female rider, the selection process created more interest than the record attempt. With the mainstream press picking up on the story, the Lambretta brand got huge media exposure. Bob is seen with the final three contenders from an initial field of 67 women candidates. At the back is long-time friend and associate Peter Baker

Maurice was now the head sales manager, a position he had been rewarded with thanks to his hard work and effort over the years. Others had vacated the position for it to happen, but Maurice stayed loyal and had gained tremendous experience over the years. A downside to the success of Lambretta Concessionaires was that those in specific departments were headhunted and enticed away to other companies. No one could blame it for happening, knowing these people had unique qualities and were bound to be in demand. Maurice was one of those people, but rather than jump ship, he was happy to be at its helm and reap the rewards on offer. However, he did not rest on his laurels for one minute, realising each day was a battle against the competition. It was a similar position Bob found himself in as he to realised he must deliver in the faith Peter Agg had in him. Perhaps it was slightly different in that the company had employed him for a much shorter time than Maurice, but he had still had to work his way up the ranks. By now, he was group publicity manager for all of Peter Agg's companies and exploits, which were added to all the time. However, it was the Lambretta that took president over everything else and this required Bob to work flat out on ideas while not forgetting he was still running the LCGB.

Bob's wife had the idea to get nice-looking people at events and rallies. On the left, presenting a J 125 to Bromley Innocents acrobatic display team. To the right, Standing next to Joan Thorn, Queen of the road 1966

They could be described as the dream team, with Peter sitting at the top and Maurice and Bob just below. Underneath them was a group of dedicated employees who all played roles in taking the company forward. Those three had high-pressure positions, and Bob remembers well how tense Peter Agg became, which was most of the time. Not a day went by where he wasn't slamming the phone down or shouting to Edna, the company secretary, to sort things. They all had their own office, but with the paper-thin dividing walls, it was pretty easy to listen in on Peter ranting about something. The same could be said of Maurice and Bob, who could hear what the other was saying, and though they got on, there was still some friction between them, mainly due to the LCGB. The members were still interfering with Maurice's work and complaining about problems regarding the Lambretta.

They may have been only minor niggles regarding faults, but that wasn't his problem; he was paid to sell them, not fix any issues; that was for the warranty department to

sort out. The problem was that Bob ran the LCGB, and Maurice would bicker with him over the matter, politely asking him to tell them to keep their noses out of his business. He just wanted to sell them, go home and forget about the job until the next day, not to have to answer questions that were nothing to do with him.

Maurice explaining the finer points of the Lambretta to John and Cynthia Lennon

However, changes were happening on that front as Bob slowly relinquished his role with the LCGB. Not that he wanted to, but he couldn't be in two places at once, and with the publicity manager's job taking up more of his time, something had to give. From now on, he handed the reigns over to Peter Davis to run the club, with Bob still promoting it and being involved where he could. There was no way he would drop it entirely as it meant too much to him, not forgetting it was him who made the club fashionable with the changes he implemented. Not only would it free up more of his time, but it also would stop Maurice from venting his fury at him regarding the club members now that he wasn't directly involved with them. Now they could focus on the more significant challenge of selling the Lambretta, and this is where Bob went into overdrive.

He had come up with many slogans in the past, which had all been successful not just while he worked at Lambretta Concessionaires but going right back to his time at Smee's. It was vitally important that adverts make the maximum impact because of the reduced budget for the department. A poor sales campaign could be devastating in terms of sales and not easy to recover from.

Olympic gold medallist Ann Packer receiving an LCGB gold medal from Bob. It was this kind of integration with famous people that was making the Lambretta look good in the 1960s

He had great success in tying the Lambretta to other companies, a policy carried out for years regarding products directly involved, such as tyre, oil and spark plug companies. Now he wanted it to be associated with famous brands even if they had nothing to do with the Lambretta. One such tie-up was with BEA (British European Airways), a BOAC division. The company was connected to the commercial launch of the Trident in 1964, hailed as the era of the Jet Set, and Bob liked the sound of it; soon enough, he negotiated a deal between the two. It allowed adverts and slogans that made the Lambretta look fashionable to gain a much wider audience, far more than an advert in the newspaper. The saying made the difference, and "join the jet set on a Lambretta" became the phrase on the tip of everyone's tongues. Bob liked it so much that Jet Set became the name for the new magazine introduced by the LCGB, which still exists today. What it showed was his thinking outside of the box in selling the Lambretta brand beyond its usual surroundings, and while it may have been commercialism, it was helping sales immensely.

Bob's tie-up with BEA and the Trident aircraft was a match made in heaven. It also introduced the Jet-Set logo that the Lambretta would become synonymous within Britain. This photo was used for advertising and taken at Gatwick airport. While models were used, Bob made sure he was involved, pictured carrying the case and wearing a trilby hat

Connecting the two brands publicly was a great idea and used to help sales of the J range, which had been struggling

It was needed as these were tough times, and taking stock of the current model line-up didn't make for good reading. The J range had been an uphill task from day one, and spending money on advertising seemed a significant waste, as Innocenti was scaling back on its production anyway. The SX 200 was the flagship model and indeed doing well, but other than that, all it left was the Li 150. That had been around for five years, so an extensive advertising campaign wasn't going to sell thousands more, and while it had been very successful by now, it was looking a bit dated. Thankfully change was on the way, and the big announcement that shocked everyone came in the news that the Li 150 would be discontinued. There was no replacement or upgrade of it just finished, but it meant Innocenti required something to take its place, which came in the way of the SX 150. Regarded as the smaller brother of the SX 200, it had a powerful motor and equally impressive performance. It also gave Bob a clear direction of which to go in when it came to advertising it.

The 1960s was the age of the sexual revolution, from the introduction of the contraceptive pill and the story of Lady Chatterley's lover to the Christine Keeler affair and the mini skirt. It may have been the early shoots of the revolution, but there was no doubt that sex was selling, and advertisers were well aware of the fact. This is where Bob came up with his eureka moment taking the Lambretta name and making it more provocative. With the SX 150 having just been launched, he came up with the idea of SX appeal and placing a full stop in the shape of a square between the letters as first glance looked like is spelt "sex appeal". There was no complaint from Peter Agg, and he was well aware of what Bob had done, even if it was on the edge of being too controversial. Though he and Maurice were given free rein to run their departments, Peter checked every letter or press release they wrote before it could be sent out. Sometimes he would want them altered, but Maurice would argue the reasons for what he had written and refuse to change it. It would usually be Peter that backed down when the whole explanation was given, as, at times, it became a war of attrition between the three men.

The SX Appeal advert was just on the accepted borderline of what was allowed but made clever use of the model's name. At the same time, the J 125 "amphiscooter" generated huge interest within the press. Bob said, "we used it time and time again until finally, one day, it sank"

They all had one goal to keep selling the Lambretta regardless of how harsh the market was becoming. For Peter and Maurice, it was a commodity, and once they had left work for the day, it was forgotten about until tomorrow. Bob had a different view because he loved the lifestyle that followed the Lambretta. It was different for him because of his connection with BLOA and the LCGB and his being the frontman mixing with customers and enjoying what it had to offer. He understood he had to deliver in terms of advertising and making sure the message got out there, but the other two were the ones that felt the strain. The problem was it wasn't just about people's attitudes to travel anymore; there were other factors causing concern with the business of selling them. Though jobs were plentiful and British manufacturing was doing well, the country was in quite a mess regarding its economy and exports. The Labour government under the leadership of Harold Wilson was in big trouble. With inflation rising and seemingly out of control, this impacted the cost of a Lambretta. Peter Agg would moan weekly about how poorly the country was being run, often shouting at Bob and Maurice regarding how much the government was costing him.

At the opening of Lambretta house in 1968, his speech mainly consisted of quotes regarding the fact. It was hardly surprising as Harold Wilson had just committed the cardinal sin of devaluing the pound. Though it may have helped exports, it put around 20% on the cost of a Lambretta overnight because that was an import. There were rumblings in Italy, with Innocenti having disputes with the workforce as the unions demanded better wages and conditions. There wasn't much the dream team could do

about it other than persevere. They hadn't achieved this success without hard work and determination, so what if it was more difficult than ever? They were all dam proud of what they had created and were not ready to throw the towel in just yet. Unfortunately, the new Lambretta that was about to be released didn't instil them with much confidence, certainly not Maurice. Branded the Luna Line, it came into Britain as a 75cc machine called the Vega, the likes of which had never been seen before. Designed by Bertone, it used the J range engine layout but reverted to the tubular chassis with parts bolted to it. Maurice hated it and thought it was flimsier than the Cento and J 125, regarding it as a backward step.

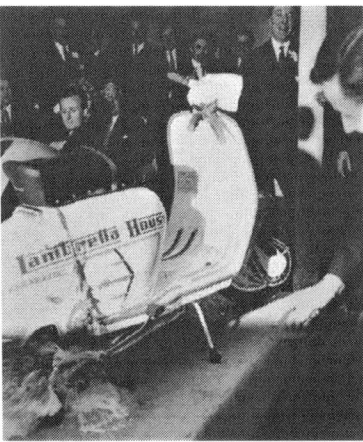

The opening of Lambretta house was a big occasion, and Peter Agg used his connections to get well-known people to attend. Among them were motorcycle Grand Prix racers Bill Ivy and Mike Hailwood, while cracking the champagne was Bruce McLaren

Visits to the Innocenti factory were frequently made, and Peter Agg invited dealers to make the journey. On the right, Peter receiving instructions before boarding the coach to the factory

On some visits, Bob would attend to carry out meetings with the LCI. They held him in great esteem, so much so that he was invited to meet the pope on one occasion. Pictured shaking his hand with his wife Ursula standing behind

He told Peter Agg that no one would buy it because sales these days were mainly to the young generation who wanted style, power, and speed. The small compact design looked futuristic and modern, but its performance was lacking compared to the SX models. Peter explained there was no choice, and thankfully both SX models would remain, but Maurice warned him that there was no way the Vega would turn sales around. For Bob, it was different as he had something new to work on. With the space race heating up as the Americans fought to land on the moon first, he had the perfect backdrop. All adverts referred to the fact, and anything from Luna to Martian was used in the advertising slogans. Appealing they may have been, but it soon became evident that no matter how good Bob was at his job, even he couldn't make a success of it.

The J range had failed to capture the public's imagination when sales were more buoyant in the middle of the decade, and with the shift of attitudes, there was no chance with the Vega. Dealers were struggling to move them in any volume, so Maurice had to come up with incentives by offering massive discounts on job lots allowing dealers to reduce their prices. Peter Agg felt like they were giving them away, and the reality was they were hardly making a profit. The problem was Innocenti was putting pressure on him to move significant quantities, but no matter how much they tried, they simply couldn't. Maurice and Bob constantly worked to develop new ideas and schemes to help the situation. The problem was that despite all their efforts, the results weren't happening and disheartening though it may have been, they had come too far to throw it all away.

Both Peter and Maurice endorsed the Luna line, but behind the scenes, they knew it would struggle more so Maurice, who had grave reservations regarding it

Innocenti was well on the way to producing a successor for the SX range; for all of them, it couldn't come quick enough. No one had any idea how much Peter Agg knew about it, but it would be naïve of Innocenti to keep him out of the loop regarding its progress. They relied on the British market taking a considerable quantity of their machines once they were in production and would try and get confirmed orders in advance. Regardless of what he thought, they were left in the dark; whether this was a ploy by Peter to keep them on their toes was anybody's guess. It did mean that once they were told, the sales machine would go into overdrive. When the new model did arrive, they were impressed, and all came to the same conclusion that it could turn their fortunes around. The Grand Prix, as Peter Agg branded it, was a breath of fresh air with its stunning looks and bright, vivid colour range. It was a clever revamp of the SX range perfectly executed by Bertone, even though there were some concerns. Maurice pointed out that while the styling was modern, the technology wasn't commenting on it being outdated.

The Slimstyle range had been a considerable success, especially once the TV 175 and 200 models had been introduced

Likewise, the introduction of the SX series in 1966 which paved the way for the Grand Prix

Of all the adverts Bob created for the Grand Prix, this one was by far the most controversial with all its double meanings

By now, the Japanese were slowly making headway into Europe, and the once-perceived threat was becoming a reality. It was more for the big motorcycle manufacturers to worry about, but that didn't mean the scooter was immune from it. The problem with the Lambretta was that the engine, brakes, electrical system, and chassis hadn't changed in almost a decade. What was being developed by the Japanese was already years ahead. While it was a concern, the new model would be good enough for now, but things would have to change in the future; they all realised that. Maurice and Bob got to work immediately, and while the orders came through thick and fast, that didn't mean there was no need to advertise it. Quite the opposite as with a good sales campaign this could set the company up well going into the next decade. At this point, Bob came up with a series of successful adverts that were probably some of the most controversial the advertising industry had ever seen.

Going off the success of the "SX appeal" logo, he came up with a series of suggestive quotes which, while referring to a sexual nature, were also true of the Grand Prix performance. They included "the virile one", "real goer", and "more poke", with each advert featuring a male rider while a provocatively dressed female looked on. The idea was heavily frowned upon by Edna Calder, Peter Agg's sectary but her objections fell on deaf ears as the series of adverts were given the go-ahead. As far as Bob was concerned, they fitted in perfectly with how society was perceived at the time, and while they were suggestive, they had a double meaning, so it depended on how the public read them. What it did give was maximum impact aimed at the young generation, who were now the majority buying the Lambretta, and there is no doubt the campaign was a huge success. Not only did they turn around the slide in sales, but they were also making the Lambretta popular again. This was the dream team at its finest, proving they still had what it takes to run a successful business. There were still dark clouds looming on the horizon, though, and it would require careful planning to keep the company going forward on a solid footing. The question remained could they as a team steer the ship through the stormy seas, and would they all survive?

Chapter Seven

The survivors

*N*othing lasts forever in this world, or if it does, it can't remain the same, and that couldn't be any truer of the Lambretta. The problem was that much of its makeup was slowly becoming outdated against the opposition. Annoyingly for Peter Agg and his team, it was out of their hands as they weren't the manufacturers; Innocenti was and, by now, in significant financial trouble compounded with strikes by their workers. The company had gotten into the situation through a catalogue of bad decisions and mistakes, and though they were trying to turn their fortunes around, it would be touch and go whether they would survive.

The styling of the Grand Prix was ultra-modern and a real head turner. The problem was that the technology that lay underneath the bodywork was somewhat outdated by the time of its release

While Peter Agg would be forever grateful for the lifestyle the Lambretta had given him, if it couldn't continue to provide, something else would have to take its place. That was easier said than done, so in the meantime, the Lambretta was the priority, and there were plans to try and expand the customer base further afield. Peter Agg had the idea to take it to Australia, which was still an unexplored market, thinking there was a vast untapped potential out there. Innocenti had tried something similar with North America but had failed, the country being too big for a small capacity machine like the Lambretta. It wasn't without trying, but by the end of the 1960s, they had virtually given up trying to break into the most significant market.

Undeterred, Peter Agg believed his team had what it took to succeed down under, almost like they were invincible, so he put his plan into motion. He proposed to Innocenti that he would set up a concession in Australia with his own funding but have the rights to sell it in that territory, similar to the deal he had agreed when the TV 200 was launched in Britain. This time it would be with the Grand Prix and Vega, to which Innocenti quickly agreed, it wasn't their money being used, and they would benefit by selling more machines. Another reason for doing it was the rising costs in Britain because of the dire situation the economy was in. The devaluation of the pound, rising purchase tax and inflation had seen the company profits continually being eaten up. It was like they were being driven out of the country through no fault of their own, and Australia, it seemed, didn't have such problems. The plans involved setting up a vast warehouse in Sydney to store the machines and the spares for servicing in readiness to take the first shipment. The man put in charge of running the operation was Peter Baker, who was responsible for the accessory department at Lambretta Concessionaires. It was a big commitment for him to move out there, but the projected plans were to capture a large proportion of Australia's two-wheeled market within six months, and he was seen as the man to do it. He had been an essential part of the successful team and was partly responsible for introducing Bob Wilkinson into the company.

From day one, there were problems, from delays getting everything set up to working out how to convince the Australian public that Lambretta was needed in their lives. The problem was that many roads were poor outside of the big cities, and people living in rural areas had tracks to their houses. A Lambretta wasn't suited to the terrain where a farmer lived, not forgetting the remoteness involved in travelling long distances. Peter Agg was confident it was the right market and was backed by Maurice and Bob, who agreed; however, the Australian public didn't, and Peter Baker was finding it hard going. His reports didn't make for good reading back in Britain, so Peter decided to

send Bob out there for a few weeks to see if he could weave some of his PR magic and improve sales. Along followed the usual stunts such as being photographed with a Kangaroo in an attempt to relate to the surroundings, but it would take much more than just that. Bob came up with the idea to do a high-profile press launch to the Australian media while putting the Lambretta through its paces. It had worked well in the past, such as the series one day a Crystal Palace racetrack and whether that was the inspiration a decade later, who knows, but a similar idea was put in place. This time it was to be at a track in Melbourne, but the circumstances were much different as Bob himself would parade around on a Grand Prix 200.

He was organising many things regarding the trip and what he required before the press day, so he didn't have time to sort the machine out. His instruction was clear to take a brand new GP 200 and get it run in correctly so he could impress the reporters with its performance. Somehow that request got lost in translation and wasn't carried out, leaving him with an untried engine. To make matters worse, Bob remembers it being a blistering hot day and not ideal for running a Lambretta flat out. He only found out about the engine's lack of running time upon arrival, leaving it too late to sort out and with a large attendance waiting to see what the flagship model could do, it was a nervy moment. Bob did his best taking it leisurely for the first few laps, probably allowing the photographers to get images at low speed before he wound on to full throttle. The high temperatures combined with the tight-fitting cylinder bore decided to have their revenge. In an instant, the engine seized solid, and with Bob failing to grab the clutch in time, the back wheel locked and spat him off and, to make matters worse, right in front of everyone.

Thankfully he only suffered a few minor scrapes and was okay, but the damage was done to the reputation of the Lambretta. Perhaps it was an omen of what was to come as sales slowed even further and the company was haemorrhaging cash. It didn't take much longer for Peter Agg to realise he had to pull the plug on the venture; after all, he had funded it, and it was better to take a loss now rather than let it get worse. Who or what was to blame became the question he wanted answering, but in truth, it was a combination of things. While the Grand Prix might struggle on the rugged terrain, the Vega would have less chance, and the deal he struck with Innocenti required him to market both models out there. The reality was it was a limited market in the first place, and maybe better due diligence and research were required to look deeper into the market figures before attempting such an audacious plan.

The main fallout came Peter Baker's way as he bore the brunt of the blame, indeed through Peter Agg's eyes. Already having an excellent job at Lambretta Concessionaires in charge of the accessory department, then going to be the head of the new venture in Australia seemed like a big promotion. Upon his return, he was told he would be running the petrol station at the front of the factory, and this was the most significant demotion that seemed possible. Bob felt very sorry for him and thought it was too harsh, sighting that it wasn't one individual's fault; the project had failed but was a combined failure. The big problem with Peter Agg was that he never let on how

things were going with Innocenti; he kept that to himself. Everyone knew they had issues with the unions and the disruption caused by what became known as "the hot autumn" in 1969. How damaging it was to Innocenti was an unknown quantity, and Peter never let on to the others, preferring to bottle it up inside.

Things were changing rapidly, and with the introduction of the Grand Prix being a bit of a disappointment from its lack of new technology, something had to give, and it did. Peter Agg had been contacted by Hambros bank concerning Suzuki GB, which was run by ex Lambretta Concessionaires employee Alan Kimber who had helped develop the Rallymaster at the beginning of the decade. The company ran out of the James factory in Birmingham and was part of the struggling AMC group. Suzuki GB was still in its infancy and in big financial trouble owing a debt to Hambros, who had all but taken it over to try and get some of their money back. The problem with the Suzuki was its limited products, making it difficult to break into the two-wheeled market, and a lack of leadership didn't help. Peter informed Maurice of the deal to take over the company and asked him what he thought about it. They were in London at the time and walking along The Strand, where it took just ten minutes to decide, the answer being yes. Peter was driven by the hunger for success and wanted to repeat what they had achieved with the Lambretta. Having tried to do so with many weird and wonderful products during the intervening years but failing to succeed. It was different this time because it was a market they knew so well and had the team to make it happen. Maurice was especially pleased because Suzuki had just introduced the Invader, which fell within the 250cc licence limit. Capable of 90mph, it was fast, modern looking and competitively priced, but more importantly, it would appeal to the market they were trying to keep hold of with the Lambretta Grand Prix.

SUZUKI
X-5 INVADER

While all of them wanted to remain loyal to the Lambretta, business was business. The Lambretta was becoming outdated, and the Japanese manufacturers were showing the way forward. Maurice, in particular, was excited about the Suzuki X-5 Invader and quickly realised they had a winning product on their hands

Issued on behalf of:

The Lambretta-Trojan Group,   By:P.Garnett Keeler MIPR.,
Lambretta House,   50 Mayfair Avenue,
Purley Way,   Worcester Park,
Croydon. CR9 4HD   Surrey.
Tel:01-686 2499   Tel:01-337 6729

N E W   S U Z U K I   M A N A G E M E N T

Complementary Marketing Operation

Peter Agg, well-known in the two-wheeler industry
for his connection with the Lambretta-Trojan
Group of Companies, of Croydon,Surrey, has been
appointed Chairman and Managing Director of
Suzuki (G.B.) Limited of Birmingham. Arrangements
are currently being made for the future marketing
of the Japanese motor cycles and mopeds to be
carried out by the Suzuki Company with sales,
spares and administrative offices transferred from
Birmingham to Croydon.

Mr.Agg, Managing Director of the Lambretta-Trojan
Group, said that his decision to accept the
Suzuki appointment had been made because it was
considered that the marketing of scooters with
motor cycles and mopeds was a complementary rather
than a competitive operation. "It has always
been recognised", he said, "that these very
different types of machines appeal to distinctive
sections of the market, although they may be sold
by the same dealers. Obviously, therefore, my
own appointment, joined by further appointments
to the Suzuki Board of other members of the
Lambretta-Trojan Group, will enable both Lambretta
and Suzuki to benefit from the experience gained
over the last 20 years in acquiring specialist
knowledge in the marketing of two-wheelers, backed
by efficient after-sales and spares service".

*The Trojan Group press release of Peter Agg becoming chairman of Suzuki GB hardly
mentioned the Lambretta*

Lambretta Concessionaires Ltd
Lambretta House Purley Way Croydon
CR9 4HD England Phone: 01 686 2499
Telex: 21138 Cables: Lambretta Croydon

N E W S L E T T E R.

TO ALL LAMBRETTA DEALERS.   No: 5/69.

Lambretta Dealers will be interested to learn that
Mr. P. James Agg has now been appointed Chairman and
Managing Director of Suzuki (G.B.) Limited. Full details
are explained in the enclosed copies of the Press Release
and the Suzuki Dealers' Newsletter.

In providing you with this information, we are anxious
to assure you that this will not affect in any way our
Lambretta marketing arrangements: these will continue
to be controlled by our present Directors and Management
who will also manage the affairs of Suzuki (G.B.) Limited.

Any immediate queries which may arise in connection with
our Sales administration should be directed to me.

Maurice E. Knight
Group Sales Manager.   24th April, 1969.

*Fearing dealers might be worried once the news got out, Maurice sent out a press release to
assure them the future of the Lambretta was safe*

The official press release that Peter Agg had become the head of Suzuki GB was issued in April 1969. It sent shockwaves throughout the industry and the Lambretta dealers, who depended heavily on them. He assured them that the Lambretta was still the main priority, but they didn't know the extent of the Innocenti's problems like he did. All Lambretta dealers were offered the Suzuki option regardless, so those who had faith in Peter and his team took the plunge offering both types of vehicles. Hambros knew what they were doing when they contacted Peter Agg, as everyone was well aware of his success, and they had their debt repaid while he acquired Suzuki GB on the cheap. It would also offset the looming Australian fiasco as both happened around the same time, but even so, to turn Suzuki into something big would require immense effort.

As usual, Bob would head the sales campaign, but sadly, this would be the last time. He wanted to move on and start a family before it was too late and required better pay; his only option was to ask for a raise. It was refused, so with a heavy heart, he decided to quit his job at Lambretta Concessionaires. It was a huge decision to make, and leaving behind all that had been achieved with the LCGB and the successful branding of the Lambretta wasn't a decision he had taken lightly. By now, it was 1970, and things were getting bad regarding supplies from Innocenti; even if Peter Agg was trying to keep it a secret, the others were now well aware of the situation.

Gone were the days when that factory would be full of thousands of machines once it was announced that production of the Lambretta at the Innocenti factory would cease

Thankfully they had Suzuki, so if things went pear-shaped with the Lambretta, at least they had that to fall back on. In May 1971, the worst news possible was broken, and their fears were confirmed as British Leyland had taken over the financially crippled Innocenti. Their first decision was to end Lambretta production not only because they saw two wheels as a thing of the past but because the Lambretta required colossal investment to take it forward, the latter being the only true statement. It was a massive blow for Peter Agg and the company as a whole, and despite all his efforts, he failed to reverse the decision. They were battlers and survivors, and the Suzuki bandwagon would go into overdrive. In that first year, they would sell just 3000 machines, but clever advertising and marketing would see that figure rise to 80,000 per annum by the early 1970s and equal on terms with Honda as they crushed the remaining British motorcycle industry. Soon after, they hired the services of a young man called Barry Sheene, who not only would they mould into a future world motorcycle champion but sell hundreds of thousands of units along the way.

Lambretta

Suzuki (Great Britain) Limited
87 Beddington Lane Croydon CRO4TD
Surrey England   Tel. 01 - 684 9456
Telex 21138 Cables: Suzuki Croydon

NEWSLETTER TO ALL LAMBRETTA DEALERS   NO. 4/72

We have many plans for Lambretta during next Season and we shall certainly continue with the present LI150 and SX200 range.

We are hopeful that by early February/March 1973, a new and even more attractive colour scheme will be available.

In the majority of cases, it will have been made abundantly clear to all Dealers - no matter what models they are selling - that the flotation of the Pound on the World money markets has made it necessary to increase prices.   Lambretta is no exception and is also affected.

Let us hope that with the various Government steps which are being taken at this moment, prices will stabilise in the near future and that further increases will not be necessary, so that we can at least settle down to a new Season with prices pegged for a substantial period.

Our new price structure, which is operative on all orders placed after the 3rd November, is attached.

In conclusion, we would like to thank our Dealers for their very excellent support during the 1971/72 Season.

We look forward to welcoming you on STAND NUMBER TWELVE at the forthcoming International Cycle & Motorcycle Show to be held at Earls Court, London, from the 8th - 18th November 1972.

3rd November 1972

After production finished, Peter Agg did everything he could to keep the Lambretta name going, even Importing the Spanish Serveta models, but he had to face the fact that the world had moved on, concentrating on Suzuki instead

So, what become of them all? Peter Agg was extremely wealthy by the time the Lambretta had finished production, earning even more with the success of Suzuki. He would embark on many exploits, including his own formula one team, for a short while, but he continued to strive for the next big thing with such ventures as Two Four Accessories. He even tried to bring the Lambretta back to Britain in 1977 when Scooters India Limited (SIL) started production of the Grand Prix after buying its rights from Innocenti. Still, it was a short-lived venture, soon realising its story belonged in the past. He continued his involvement with historic racing cars right up to his death in 2012.

In 1974 the Trojan formula one team entered eight races. Their driver was Australian Tim Schenken, who achieved a best place finish of 10th in Belgium. Though it was only a short-lived venture, Peter Agg lived out his dream of being involved in the sport

For Maurice, he remained by Peter Agg's side and was a significant part of the Suzuki story until, in 1982, when he retired, turning his back on the industry but thankful he had been part of it. Many who had worked at Lambretta Concessionaires carried on at Suzuki GB, playing an essential role in making the company tick. Others like Pete Meads and Tony Tessier ventured off into something different, all being successful. It was as if working with the Lambretta somehow gave them an apprenticeship in life that would make them good at anything else they tried.

Then there was Bob, the charismatic character that brought the Lambretta brand up to date in the fast pace of the 1960s. He had left before the Lambretta had finished, and for whatever reason, it did him no harm. From his job with Peter Agg, he moved to a company called Staflex International before setting up his own advertising agency, Interact Consultants limited. Who should he then work with, none other than Peter Baker, the man who had been his friend throughout their time at Lambretta Concessionaires. Peter was now high up at Stihl, the chainsaw company and used Bob's services much to the annoyance of Peter Agg, who was in direct competition as he too had his own agricultural equipment company.

The story must end at the beginning with Les Ashton, who had been an entrepreneur of just about anything. He may have been annoyed somewhat at being left out of the Lambretta story by Peter Agg, but thanks to his family's commitment to fighting for the truth, he can be remembered for making it all happen in the first place. There is no doubt that Lambretta has an exceptional quality that somehow affects anyone who has ever ventured near one and still exists to this day. Perhaps that is why its appreciation has never faded and will always remain, not just with the owners. Those that brought it to us and built its legacy will never forget it either, and they should not be forgotten for making it happen; they are the Lambretta people.

Printed in Great Britain
by Amazon

85682633R00056